SLUDGE PARASITES AND OTHER PATHOGENS

ELLIS HORWOOD SERIES IN WATER AND WASTEWATER TECHNOLOGY

Series Editor: MICHAEL WINKLER, University of Surrey

WATER POLLUTION BIOLOGY
P. D. ABEL, Sunderland Polytechnic
ALKALINITY–pH CHANGES WITH TEMPERATURE FOR WATERS IN INDUSTRIAL SYSTEMS
A. G. D. EMERSON, Independent Consulting Technologist
CHEMICAL PROCESSES IN WASTE WATER TREATMENT
W. J. EILBECK, University College of North Wales, and
G. MATTOCK, Waste Water Treatment and Resource Recovery Consultant
BIOLOGICAL TREATMENT OF SEWAGE BY THE ACTIVATED SLUDGE PROCESS
K. HÄNEL, Research Leader of Purification of Sewage Team, Leipzig, Germany
ANALYSIS OF SURFACE WATERS
H. HELLMANN, Federal German Institute of Hydrology, Koblenz
WATER-ABSTRACTION, STORAGE, TREATMENT, DISTRIBUTION
J. JEFFERY, General Manager, North Surrey Water Company
CONTROL OF EUTROPOHICATION IN INLAND WATERS
H. Klapper, Magdeburg Research Department of Water Protection, Germany
POLLUTION CONTROL AND CONSERVATION
Editor: M. KOVACS, Hungarian Academy of Sciences
WATER SAMPLING
J. KRAJČA, Research Institute of Geological Engineering, Brno, Czechoslovakia
GROUND RESOURCES AND YIELDS
H. KRIZ, Institute of Geography of the Czechoslovak Academy of Sciences, Czechoslovakia
PREVENTION OF CORROSION AND SCALING IN WATER SUPPLY SYSTEMS
L. LEGRAND, Former Head of the Water Dept. of City of Paris, and P. LEROY, Chemical Engineer of City of Paris, Chief of Corrosion Laboratory
SLUDGE PARASITES AND OTHER PATHOGENS
R. LEWIS-JONES, Consultant and M. WINKLER, University of Surrey
HANDBOOK OF WATER PURIFICATION Second Edition
W. LORCH, The Lorch Foundation
SAMPLING STRATEGIES FOR MICROBIOLOGICAL ANALYSIS IN WATER DISTRIBUTION SYSTEMS
A. MAUL, D. VAGOST and J. C. BLOCK
BIOLOGY OF SEWAGE TREATMENT AND WATER POLLUTION CONTROL
K. MUDRACK, University of Hanover, FRG, and S. KUNST, Institute for Sanitary Engineering, FRG
DRINKING WATER MATERIALS: Field Observations and Methods of Investigation
D. SCHOENEN and H. F. SCHÖLER, Hygiene-Institut der Universität, Bonn, FRG
HANDBOOK OF LIMNOLOGY
J. SCHWOERBEL, University of Freiburg, West Germany
BIOTECHNOLOGY OF WASTE TREATMENT AND EXPLOITATION
Editors: J. M. SIDWICK and R. S. HOLDOM, Watson Hawksley, Buckinghamshire
MICROBIAL INTERACTIONS WITH CHEMICAL WATER POLLUTION
D. TOTII and D. TOMASOVICOVA, Institute of Experimental Biology & Ecology, Bratislava:
Edited by M. A. Winkler, University of Surrey
BIOLOGICAL TREATMENT OF WASTEWATER, 2nd Edition
M. WINKLER, University of Surrey

Titles published in collaboration with the
WATER RESEARCH CENTRE, UK

AQUALINE THESAURUS 2
Editor: G. BOTHAMLEY, Water Research Centre
BIOLOGICAL FLUIDISED BED TREATMENT OF WATER AND WASTEWATER
Editors: P. F. COOPER, Water Research Centre, and B. ATKINSON, UMIST
ENVIRONMENTAL TOXICOLOGY: Organic Pollutants
J. K. FAWELL and S. HUNT, Water Research Centre
ENVIRONMENTAL PROTECTION: Standards, Compliance and Costs
Editor: T. J. LACK, Water Research Centre
WATER RESEARCH TOPICS
Editor: I. M. LAMONT, Water Research Centre
SMALL WATER POLLUTION CONTROL WORKS: Design and Practice
E. H. NICOLL, Scottish Development Department, Edinburgh
RIVER POLLUTION CONTROL
Editor: M. J. STIFF, Water Research Centre

ELLIS HORWOOD BOOKS IN AQUACULTURE AND FISHERIES SUPPORT

Series Editor: Dr L. M. LAIRD, University of Aberdeen

ECOTOXICOLOGY AND THE MARINE ENVIRONMENT
Editors: P. D. ABEL and V. AXIAK
BACTERIAL FISH PATHOGENS: Disease in Farmed and Wild Fish
B. AUSTIN and D. AUSTIN, Heriot-Watt University
MICROBIAL BIOTECHNOLOGY: Freshwater and Marine Environments
B. AUSTIN and D. AUSTIN, Heriot-Watt University
METHODS FOR THE MICROBIOLOGICAL EXAMINATION OF FISH AND SHELLFISH
B. AUSTIN and D. AUSTIN, Heriot-Watt University
AQUACULTURE
G. BARNABÉ, Université des Sciences et Techniques du Languedoc
AUTOMATED BIOMONITORING
D. GRUBER and J. DIAMOND, Biological Monitoring Inc., Virginia, USA
SALMON AND TROUT FARMING
L. M. LAIRD and T. NEEDHAM, University of Aberdeen
PRINCIPLES OF FISH NUTRITION
W. STEFFENS, Institute of Inland Fisheries, Berlin, and Humboldt University, Berlin

SLUDGE PARASITES AND OTHER PATHOGENS

ROBERT LEWIS-JONES Ph.D.
Consultant, Tonyrefail, Mid Glamorgan
MICHAEL WINKLER
Department of Chemical and Process Engineering
University of Surrey

ELLIS HORWOOD
NEW YORK LONDON TORONTO SYDNEY TOKYO SINGAPORE

First published in 1991 by
ELLIS HORWOOD LIMITED
Market Cross House, Cooper Street,
Chichester, West Sussex, PO19 1EB, England

A division of
Simon & Schuster International Group
A Paramount Communications Company

Typeset in Times by Ellis Horwood Limited
Printed and bound in Great Britain
by Hartnolls, Bodmin, Cornwall

British Library Cataloguing in Publication Data

Lewis-Jones, Robert
Sludge parasites and other pathogens: Health risks from sewage. —
(Ellis Horwood series in water and wastewater technology)
I. Title. II. Winkler, Michael. III. Series.
628.4

ISBN 0–13–963703–6

Library of Congress Cataloging-in-Publication Data

Lewis-Jones, Robert, 1953–
Sludge parasites and other pathogens / Robert Lewis-Jones, Michael Winkler.
p. cm. — (Ellis Horwood series in water and wastewater technology)
Includes bibliographical references and index.
ISBN 0–13–963703–6
1. Sewage sludge — Health aspects. 2. Sewage — Microbiology.
I. Winkler, M. A. (Michael A.), 1935– . II. Title. III. Series.
RA567.L48 1991
616.9–dc20

91–2210
CIP

Table of contents

Preface

The aim of this book is to give a concise account of microbial pathogens found in sewage sludge and methods of sludge treatment available for inactivating them. Wastewater treatment has the effect of separating sewage and other wastewaters into purified water and a 'pollution concentrate', the sludge. The production of purified water has been the principal objective of wastewater treatment, with the resultant sludge as a waste product in all senses of the phrase. Disposal of sludge has always been a problem, and accounts for roughly half the total costs of sewage treatment. The sludge disposal problem is currently being exacerbated by an increase in the amount of sludge needing disposal and increasing severity of the regulations governing sludge disposal. Environmentally acceptable treatment and disposal of sludge must now be an equally important objective of wastewater treatment.

The UK generates about a million tonnes of sludge solids per year, of which about 30% is discharged into the sea. Dumping of sludge at sea is due to be phased out by 1998, so alternative disposal methods will have to be found. In addition, sewage from 17% of the population in the UK is discharged to sea untreated. When this practice is stopped, conventional treatment of this sewage could generate an additional 180 000 tonnes of sludge annually. Overall, the amount of sludge requiring land-based treatment will double. Environmentally acceptable treatment and disposal of sludge is thus a major problem requiring urgent attention.

Sludge contains useful amounts of nitrogen and phosphorus, and it is logical to utilize these as agricultural fertilizer. The agricultural use of sewage sludge is, however, limited by the microbial pathogens and heavy metals found in sludge. The heavy metals in sludge are beyond the scope of this book. In considering microbial pathogens and the effectiveness of various sludge treatments in inactivating them, particular attention has been given to microbial parasites, with a chapter each on flatworms, roundworms and protozoa, as these organisms are usually the least familiar to environmental technologists. In particular, the pathogenic protozoon *Cryptosporidium* is currently a matter of serious concern, as it has proved to be unaffected by standard methods of water treatment (as opposed to *wastewater* treatment). Bacteria are the most obvious pathogens in sewage sludge, and particular attention has been given to *Campylobacter* and *Salmonella* and their susceptibility to sludge treatments. In considering the effects of sludge treatment on viruses in sludge, the problem has been exacerbated by the difficulty of assaying viruses in general and enteric viruses in particular. A brief account of sewage treatment processes which generate sludge has been included, pointing out, where possible, aspects which can alleviate the problems of treatment and disposal of the sludge produced.

ACKNOWLEDGEMENT

The authors wish to express their gratitude to the World Bank, Washington DC, for their good wishes for this book and permission to reproduce diagrams from that comprehensive volume on environmental health *Sanitation and Disease, Health Aspects of Excreta and Wastewater Management,* by Richard Feachem, David Bradley, Hemda Garelick and Duncan Mara (published for the World Bank in 1983 by John Wiley & Sons).

Robert Lewis-Jones
Michael Winkler

1

Sludge disposal

1.1 INTRODUCTION

Sludge is currently a matter of urgent international concern. The quantity of sludge, defined as 'residual sludge from sewage plants treating domestic or urban waste waters', being generated is steadily increasing, while the restrictions on its disposal are steadily becoming tighter. The increase in the volume of sludge produced is partly because of the general trend of population increase but, particularly in Europe, is also due to centralized sewage treatment facilities being made available to an increasing number of people. In addition, the development of intensive animal husbandry is giving rise to increasing amounts of animal slurry. At the same time, the means of disposing of sewage and sludge is subject to an increasing amount of restriction and control. Disposal of sludge and untreated sewage to sea is being phased out in Europe during the 1990s. Sewage that was hitherto discharged to sea will now have to be treated, and will generate sludge requiring disposal and, it is to be hoped, treatment.

1.1.1 Wider implications

In addition to an expected improvement in the microbiological quality of bathing water and fisheries, it has been suggested by one of the present authors that the discontinuation of disposal of sewage and sludge at sea could have beneficial results in terms of other current environmental concerns, the stability of the stratospheric ozone layer and the so-called 'greenhouse effect' (Winkler & Manoranjan 1989). This suggestion was made on the basis that the microbiological degradation of the nitrogen content of biomass in sewage and sludge can produce significant quantities of nitrous oxide in oceanic conditions. Indeed, a survey of the Indian Ocean completed in 1989 found that the nitrous oxide content of the water was ten times greater than expected. Nitrous oxide is an efficient absorber of infra-red radiation, and its 'greenhouse' effect is nearly 300 times greater than that of carbon dioxide. In addition, it can be converted photochemically to nitric oxide, which catalyses ozone destruction (Winkler 1984). The chemical aspects of sewage and sludge disposal are, however, beyond the scope of the present discussion.

1.2 SLUDGE PATHOGENS

The restriction on disposal at sea is inevitably putting pressure on other methods of sewage treatment and sludge disposal, and, as incineration also has implications of atmospheric pollution and problems with ash disposal, the most promising method is disposal on to land. The disposal of sewage sludge on to land as a soil conditioner and fertilizer has obvious benefits, as the nitrogen, phosphorus and organic matter in sewage sludge can supply a large part of the requirements of most crops, but immediately arouses concern about its potential for causing pollution. Human and animal health risks associated with the disposal of sewage sludges to land are directly related to their toxic metal content and pathogen levels, especially when there is a lack of acceptable control criteria (Hudson & Fennell 1980), although this present review concentrates on the pathogen criteria. Any animal disease hazard could be avoided by spreading sludge on arable land, rather than on grassland (Argent *et al.* 1981), although this would create a need for greater sludge storage capacity as a result of the consequently seasonal nature of sludge disposal. Disposal on to land with a feasible uptake of nutrients by crops has been considered as an economically advantageous method of dealing with municipal wastewater (Young & Carlson 1975), provided sufficient land is available, because physical–chemical processes require skilled operators, chemicals and a considerable energy input.

Sewage sludge and wastewater contain a wide spectrum of bacterial, viral and parasitic pathogens. The range of wastewater pathogens is well-documented (Akin *et al.* 1977, Carrington 1978, Graham 1981, Reimers *et al.* 1982, Snowdon *et al.* 1989a, Abel 1989), but the study of factors affecting the survival rate of pathogens in sludge and effluent has been less systematic. This study aims to review current knowledge of parasites and other microbial pathogens in wastewater, to consider the environmental and process factors encountered by pathogens in sewage and sludge, and to evaluate pathogen survival in respect of the factors encountered.

There is thus a risk of introducing pathogens into the food chain or water supply in land disposal of human or animal wastes. The most obvious mechanisms of contamination are surface run-off, percolation into groundwater, leakage from the wastewater storage system and survival of microbial pathogens on the surface of food plants. In addition, aerosols may be created during pumping, mixing, transport and distribution of the wastewater and insects and wild animals may also act as vectors of disease transmission. It is therefore important that pathogens be inactivated to as large an extent as possible before land disposal of wastes.

Conventional wastewater treatment systems are successful in removing the majority of pathogens listed in Table 1 (Lund 1975, Feachem *et al.* 1980, 1983, Snowdon *et al.* 1989a), but there is nevertheless a need for chemical and process methods capable of destroying pathogens. This review therefore both considers the success of conventional treatment methods, such as sedimentation and anaerobic digestion, in removing pathogens and evaluates the potential of chemical and environmental factors capable of destroying or inhibiting pathogens in sludge and effluent.

Table 1.1 — Principal pathogenic micro-organisms in human and animal waste

Viruses	Bacteria	Parasites
Enteroviruses	*Bacillus*	Protozoa
Poliovirus	*Brucella*	*Cryptosporidium*
Coxsackieviruses A & B	*Campylobacter*	*Entamoeba*
Echovirus	*Clostridium*	*Giardia*
Hepatitis A	*Enterobacter*	*Toxoplasma*
Adenovirus	*Escherichia coli*	Nematodes (roundworms)
Reovirus	*Leptospira*	*Ancylostoma*
Rotavirus	*Listeria*	*Ascaris*
	Mycobacterium	*Necator*
	Pseudomonas	*Strongyloides*
	Salmonella	*Toxocara*
	Shigella	*Trichuris*
	Staphylococcus	Cestodes (tapeworms)
	Yersinia	*Echinococcus*
		Taenia

Many epidemics of water-borne diseases occurred before the pathogenic nature of micro-organisms was known. Snow (1855) was the first scientist to link water and disease when he demonstrated that cholera was transmitted by water contaminated by faeces. It has since been said that treated water supplies were infrequent sources of enteric disease because of improvements in water treatment processes and the procedure for the detection and improvement of pollution indicators (Kabler *et al.* 1963). Thcsc improvements are considerable in the developed western world, but many tropical and Third-World countries experience poor sanitation and financial difficulties which result in an abundance of wastewater-associated disease (Bradley 1977, Evision & James 1977, Dahling *et al.* 1989, Black *et al.* 1989). Although the principal water-related diseases are considered briefly, this present review is concerned mostly with pathogens and parasites in sludge, and their significance in respect of the use of sewage sludge as a fertilizer for agricultural land.

1.3 THE SLUDGE PROBLEM

The annual production of sewage sludge in the UK is about 30 million tonnes of wet sludge, containing about 1.2 million tonnes of dry solids, which cost about £200 million for treatment and disposal at 1986 values (Davis 1987). The EC produces annually about 5.6 million tonnes (as dry solids) of sewage sludge, of which, in 1984, 37% was recycled to land in agriculture, land reclamation, horticulture, forestry and parkland , 44% disposed into landfill sites, 9% incinerated and 7% disposed of to sea. Prior to disposal, 79% of the sludge was treated, mainly by mesophilic anaerobic digestion or ambient aerobic digestion, and 71% of the sludge was de-watered (Newman *et al.* 1989).

On average, some 60 g of sludge dry solids are disposed of each day for every EC citizen connected to sewerage. The UK was the second largest sludge producer in the EC, producing over a million tonnes of sludge dry solids in 1984, of which about 41% was used in agriculture and 13% in land reclamation and as landfill. In spite of this apparently high proportion of agricultural utilization, only about 1–2% of agricultural land in England and Wales actually receives sludge, which accounts for about 1.2% of the nitrogen input and 2.2% of the phosphorus input (Davis 1989).

Sludge disposal figures for 1986 were cited by Brunner and Lichtensteiger (1989) in reviewing the practice and legislation in Europe with regard to the use of sludge for landfilling. In 1986, 350 million people in 15 European countries gave rise to 6 million tonnes of sewage sludge (dry solids), of which 43% (2.6 million tonnes) was disposed of in landfills. In the UK, only 9% was disposed into landfills, of which 51% was both stabilized and de-watered before disposal, 39% was de-watered but not stabilized, 2% was stabilized but not de-watered and 8% was untreated. In many European countries, increasing use of landfill disposal results from the decreasing use of sewage sludge in agriculture and, possibly, from its being less closely regulated than any other disposal method. Most countries have no national regulations at all relating to landfill disposal of sewage sludge, although some are licensed locally, while others have regulations relating only to the geological stability of the landfill.

An increased demand for land disposal may, however, be expected, as discharge of UK sewage sludge to sea, which accounted for 30% of sludge disposal in 1984, is due to be phased out by 1998. In addition, raw sewage is discharged into the sea by more than 500 pipelines from the UK, and this will now have to be treated (Brown 1990). The sewage from about 17% of the population is discharged to sea and estuarial outfalls, which will produce an estimated additional 180 000 t of sludge dry solids if subjected to conventional land-based sewage treatment. For comparison, it has been estimated that the yearly production of municipal sewage sludge in the USA is 9 million tonnes (Cunningham & Skinwood 1990), and municipal sewage discharged into the coastal waters of the United States amounts to about 30 million tons, (8 billion US gallons, 3×10^{10} l) of which only half receives secondary treatment (Gerba & Goyal 1988).

While UK sludge production was predicted to increase by only a few per cent in 1994, for the EC as a whole, the increase predicted was 33%, to over 7 million tonnes, of which 37% will be recycled to land and 47% to landfill sites (Newman *et al.* 1989). Some of the predicted increase results from an increase in the number of people connected to public sewers, from 86% in 1984 to a predicted 88% in 1994. Recent estimates of municipal sewage sludge production in Canada and the USA are 0.422 and 9 million tonnes dry solids per year respectively (Cunningham & Skinwood 1990).

1.4 THE EC DIRECTIVE

There is clearly considerable current concern regarding the potentially pathogenic nature of sewage sludges and their use as an agricultural fertilizer (Council of the European Communities 1986, Ottolenghi *et al.* 1987, Department of the Environment 1989, Davis 1989, Snowdon *et al.* 1989a, Hall 1989b, Owen & Stansfield 1989, for example). The European Commission initiated a research programme on sewage sludge in 1971, under five headings: processing, chemical contamination, hygienic effects, agricultural value and environmental effects. The Commission's 1986 directive 'on the protection of the environment, and in particular the soil, when sewage sludge is used in agriculture' was formulated after noting the reports of the five working parties, with two main objectives (Newman *et al.* 1989). The first objective was to ensure that human beings, animals, plants and the environment in general are fully safeguarded against possible harmful effects of uncontrolled disposal of sewage sludge on agricultural land. The second was to promote the correct use of sewage sludge on land, possibly as an initial stage in the formulation of a Community policy on soil protection.

The directive recognizes that sludge can have valuable agronomic properties and its use should be encouraged, provided it is used correctly. In Article 2 of the directive, the term 'sludge' is taken to mean the residual sludge from sewage treatment plants, including those dealing with wastewaters of domestic or urban origin or with a similar composition and from septic tanks and similar installations. The EC directive specifies that sludge is to be treated before being used in agriculture, and defines treated sludge as sludge that has had its fermentability and health hazards resulting from its use reduced by undergoing an appropriate process, such as biological, chemical and/or heat treatment, or long-term storage. 'Use in agriculture' then means the spreading or any other application of sludge on or in the soil, in connection with the growing of all types of commercial food crops, including those used for stock-rearing purposes. Much of the directive, Articles 4, 5 and 8 and Annexes 1A, 1B and 1C, are concerned with concentrations of specified heavy metals (Cd, Cu, Ni, Pb, Zn, Hg and Cr) in soil and sludge.

Article 6 specifies that sludge shall be treated before being used in agriculture, although the use of untreated sludge may be authorized provided it is injected or worked into the soil. According to Article 7, the use of sludge is to be prohibited on soil in which fruit and vegetable crops are growing, apart from fruit trees, on grassland or forage crops if the grassland is to be grazed or the crops harvested before at least three weeks have elapsed, and during the harvest, and for ten months preceding the harvest, of fruit and vegetable crops in contact with the soil and normally eaten raw. Article 8 then specifies that the quality of the soil and surface and ground water is not impaired by the use of sludge, and that sludge is used in accordance with the nutrient requirements of plants. Article 8 also specifies that account shall be taken of the increased availability of heavy metals to plants when the soil pH is below 6.

1.4.1 Odour

In respect of the reduction in sludge fermentability specified by Article 2(b) of the EC directive, Davis (1989) points out that sludge odour is a common cause of complaint, and common-sense procedures have been formulated in the code of practice in respect of the means of distributing sludge over the land and the route taken in transporting sludge to the site of application. Avoidance of surface and groundwater pollution is, of course, essential, and account must be taken of the proximity of the site of application to waterways and water sources, as well as the rate of sludge application, the soil conditions and the topography of the area. Recent developments in methods of applying sewage sludge to land have been reviewed by

Hall (1989a), and odour problems with sewage sludge, and some approaches to quantifying and controlling them, by Voorburg & van den Berg (1989).

1.5 SLUDGE REGULATIONS IN THE UK

The EC directive is also aimed at harmonizing guidelines for sludge utilization among EC member states, and the EC directive was implemented in the UK by The Sludge (Use in Agriculture) Regulations 1989, which came into force on 1st September 1989 (Statutory Instrument 1989 No. 1263). The wording of the regulations is very similar to that of the EC directive, except that use of sludge on soil with a pH value less than 5 is prohibited. This is presumably to limit the mobility of heavy metals, although low soil pH is an advantage in terms of inhibiting the survival of some pathogens. Provided the sludge is worked into the soil as soon as reasonably practicable, the regulations allow untreated sludge to be used on land without being injected into the soil. Special provision is made for agricultural land used exclusively for sludge disposal, called 'dedicated sites', where only crops for animal consumption may be grown.

The constraints on grazing and harvesting specified in the 1989 UK Regulations (Statutory Instrument 1989 No. 1263) prohibit the use of sludge when fruit or vegetable crops are growing or being harvested in the soil, except for fruit trees, and fruit and vegetable crops grown in contact with the ground and normally eaten raw must not be harvested for ten months after the use of sludge. Forage crops are not to be harvested, nor animals grazed, for three weeks after the use of sludge. Untreated sludge must be worked into the land as soon as reasonably practicable.

In a review of the implications of these regulations for sludge treatment and disposal, Bruce *et al.* (1990) conclude that, in general, existing disposal and treatment practices already conform to the provisions of the new regulations, although modifications of sludge processing at sewage treatment works and/or the rate of sludge application to land may be necessary. More sampling and analysis of soils, and registration of information, will be required. In particular, sludge from all sewage treatment works, including those serving small populations, will be subject to the new regulations. It is pointed out by Bruce *et al.* that humus sludge, surplus activated sludge and sludge from extended aeration treatment are not considered to be biologically treated.

1.6 UK CODES OF PRACTICE

Guidelines on sludge treatments meeting the UK regulations are given in the *Code of Practice for Agricultural Use of Sewage Sludge* (Department of the Environment 1989). These were formulated to ensure that the use of sludge in agriculture does not conflict with good agricultural practice or put human, animal or plant health at risk, that water pollution and other public nuisances are avoided and that the long-term viability of agricultural activities is maintained. These guidelines are clearly in harmony with the EC directive. A companion publication, *Sewage Sludge in Agriculture — Environmental Costs and Benefits,* is in preparation, discussing the scientific basis for the code of practice, but is not available at the time of writing.

1.6.1 Benefits of using sludge in agriculture

In considering the benefits obtainable from the use of sewage sludge in agriculture, the 1989 *Code of Practice for Agricultural Use of Sewage Sludge* noted that the nitrogen, phosphorus and organic matter present in significant proportions in sewage sludge can supply a large part of the requirements of most crops. Sludge, defined as 'residual sludge from sewage plants treating domestic or urban waste waters', should, however, be applied at a rate consistent with the needs of the crops at a time as close as practicable to when the crops can utilize the sludge nutrients. In the year of application, about half the phosphorus content of sewage sludge is available to plants, irrespective of the sludge treatment used. The sludge treatment does, however, affect the availability of the nitrogen content of sludge. The high content of ammonia-nitrogen in liquid anaerobically digested sludge is readily available to plants and particularly beneficial to grassland. The release of nitrogen from untreated sludge and de-watered treated sludge is much slower, so that the benefit to crops takes longer to be realized. The organic solids content of sludge acts as a useful soil conditioner, particularly de-watered sludge cake.

1.6.2 Microbial pathogens in sludge

There is naturally considerable concern over the possibility of disease transmission from spreading sewage sludges on land. Sewage inevitably contains certain pathogens, depending on the sources of effluents in the catchment area of the sewage treatment operation and the general state of health of the community served. A WHO working group in 1981 drew particular attention to four pathogens found in sludges, *Taenia saginata, Salmonella, Sarcocystis* and hepatitis A. Of the organisms found in sludge, among those of most concern in the UK are *Salmonellae,* eggs of the beef tapeworm *Taenia saginata* and its larval stage in cattle *Cysticercus bovis,* potato cyst nematodes *Globodera* spp. and a range of viruses. The main transmission routes in food animals are directly from animal to animal or indirectly through food and animal slurry, and *Salmonella* in man through faulty hygiene in food preparation, usually in handling meat and dairy products (Davis 1989). There are, however, other possibilities, such as defaecation on to land and transmission by birds from sewage-handling operations.

In a review of the disposal of sludge to land in 1981, the UK Department of the Environment and the National Water Council identified four groups of pathogens as potential sources of infection in the UK. Only *Taenia saginata* was cited as definitely being disseminated through the disposal of sewage sludge, but ova of other parasites, *Taenia solium, Ascaris* and *Trichuris* were a cause for concern. Among bacteria in general, *Salmonella* was specifically mentioned, enteroviruses among viruses in general, and *Giardia* among protozoan cysts.

It was recognized by the UK North-West Water Authority in their 1983 code of practice that the two types of organism providing a major disease risk through the land application of sewage sludge are the *Salmonella* group bacteria and the beef tapeworm *Taenia saginata.* The rate of sludge application should be determined by the solids concentration and the content of nutrients, metals and pathogens in the sludge. The use of the land and the composition and nature of the soil must be taken into account, as must the method of application of sludge and the degree of control of the operation. The number of applications per year, and restrictions caused by seasonal weather changes, and the available sludge storage capacity must also be considered. These are implicit in the 1989 regulations governing the use of sewage sludge in agriculture (Statutory Instrument 1989 No. 1263, The Sludge (Use in Agriculture) Regulations 1989), although the application rates, specified in regulation 3(4) are determined by amounts of heavy metals in the sludge and in the soil. The Department of the Environment 1989 Code of Practice specifies additional sludge parameters, including the pH and the content of dry and organic matter, total and ammoniacal nitrogen, phosphorus and additional elements including arsenic and fluoride.

The 1989 Code of Practice recognizes that sludge contains pathogenic bacteria, viruses, protozoa and other parasites constituting a potential health hazard to humans, animals and plants. The organisms considered to be of most concern are *Salmonella,* the eggs of the beef tapeworm *Taenia saginata,* potato cyst nematodes and a number of viruses. This potential health hazard can, however, be substantially reduced by treatment of the sludge before land application, with further reductions effected by soil micro-organisms and the weather. The North-West Water Authority, in their 1983 code of practice put acceptable limits for *Salmonella* at 100 organisms per 100 ml sludge where crops are to be eaten raw, but 2400 *Salmonella* organisms per 100 ml when crops are to be cooked before eating. The UK Department of the Environment 1989 Code of Practice recognizes that it is not practicable to express the microbiological quality of sludge in terms of numerical limits for routine monitoring purposes. It does, however, warn that the microbiological quality of treated sludge needs to be checked periodically when quantities of wastes from animal or poultry processing are discharged to sewers. It also specifies that specialist advice should be taken before the use of sludge from wastewater containing effluent from the processing of hides from countries where anthrax is endemic (see, for example, section 9.8).

The 1989 Code of Practice precludes the use of sludge on growing fruit and vegetable crops, or under cover, and the use of untreated sludge in orchards or on land used for nursery stock. It lists examples of acceptable uses of treated sludge in agriculture, and additionally specifies that the ten-month rule for fruit eaten raw should apply to fruit trees, but sludge should not be applied to land used for seed potatoes or nursery stock. Similar guidelines are given for injection of untreated sludge.

The code of practice on sludge utilization actually lists examples of suitable sludge-treatment processes and operating conditions, and recommends suitable practices for land use in respect of pasture and fruit and vegetable crops which may be eaten raw (Davis 1989). Examples are given in section 2.4.

1.7 SEWAGE REGULATIONS

1.7.1 UK regulations

Clearly, the composition of sewage sludge depends on the constituents of the sewage entering the sewage treatment system. A brief account of the regulations governing discharges to sewers in the UK is therefore given here. The regulations have been built up as the result of a series of Acts of Parliament, most recently the Control of

Pollution Act 1974 and the Water Act 1989, to which will be added the Act arising from the Environmental Protection bill going through Parliament at the time of writing.

1.7.1.1 Discharge consents

The principal method of control of polluting discharges is by means of *consents*, agreed between the discharger and the appropriate authority. Authorization for consents, other than for discharge to sewers, is the responsibility of the National Rivers Authority (NRA). Authorization for consents for discharges to sewers are, with certain exceptions, the responsibility of the water company dealing with the sewage. The exceptions relate to discharges which involve substances and/or processes prescribed under the Water Act 1989 or under the Environmental Protection bill. Consents are the responsibility of the Secretary of State in the first case and of Her Majesty's Inspectorate of Pollution (HMIP) in the second case. A water company cannot authorize consents in these two cases, but it can *refuse* consents in both these as well as in cases for which it is normally responsible.

1.7.1.2 Integrated pollution control (IPC)

HMIP will be made up of several regulatory bodies formerly responsible for the separate control of waste disposal, water pollution and air pollution, to operate a new system of **Integrated Pollution Control** (IPC). Authorization of the enforcing authority, either HMIP or, in respect of certain types of air pollution, the Local Authority, will then be required for operation of a prescribed process, and IPC is expected to apply to about 3500 industrial sites. The process operator will be required to minimize the release of prescribed substances by using 'the best available techniques not entailing excessive cost' (BATNEEC). The conditions imposed will also aim at 'the best practicable environmental option' (BPEO) causing the least damage to all environmental media, air, land or water, as a whole, on both a short- and long-term basis.

1.7.1.3 Prescribed processes and substances

The UK Department of the Environment has nominated a number of hazardous substances whose discharge to water should be minimized, known as the **Red List**, and industries discharging Red List substances should then be subject to emission standards. The regulations formulated under Part I of the Environmental Protection bill establishes a list of processes scheduled for IPC to control major discharges of Red List substances or large amounts of other special wastes. The substances prescribed for water include cadmium and mercury and their respective compounds, certain chlorinated hydrocarbons and phenols, polychlorinated biphenyls, organotin compounds and certain pesticides. Prescribed processes include the production of chlorinated organic chemicals, paper pulp and asbestos products (Statutory Instruments 1989, No. 1156, Water England and Wales, The Trade Effluents (Prescribed Processes and Substances) Regulations 1989, coming into force 1st September 1989, and 1990, No. 1629, Water England and Wales, The Trade Effluents (Prescribed Processes and Substances) (Amendment) Regulations 1990, coming into force 31st August 1990).

1.7.2 EC regulations

1.7.2.1 Regulation of discharges

The EC Directive 76/464 of 4th May 1976, on pollution caused by certain dangerous substances discharged into the aquatic environment of the Community, establishes the general principles of the control of effluent discharges to waterways. The Directive established two lists of substances according to their polluting effects, with the aim of eliminating pollution by substances on List I, the 'Black List', and reduce pollution by those on List II, the 'Grey List'. Subsequent directives then set standards for different types of receiving waters and for specific substances.

Black List substances are toxic and/or tend to persist and accumulate in the environment, such as cadmium and mercury and their respective compounds, organo-halogen, -tin and -phosphorus compounds, persistent hydrocarbons and proven carcinogens. Grey List substances have deleterious but containable effects on the aquatic environment, such as less persistent hydrocarbons, specified heavy metals, silicon and phosphorous compounds, biocides, cyanides, fluorides, nitrites and ammonia.

1.7.2.2 Sewage treatment standards

A draft directive on the treatment of municipal wastewater, COM(89) 518 (final), with the aim of protecting the environment has also been issued by the EC. This specifies discharge requirements for municipal wastewater treatment plants, and its implementation will have a considerable effect on the treatment of wastewater in Europe. EC member states must compile a programme for implementing the directive by 1992 and comply with it by 1998. Treatment plants will have to reduce the wastewater BOD (biologically determined oxygen demand) by 70 to 90% to give a maximum daily average BOD of 25 mg/l in the treated effluent, a maximum daily average COD (chemically determined oxygen demand) of 100 mg/l and suspended solids of 30 mg/l. Nitrogen and phosphorus levels must also be reduced by 80%, to give a maximum daily average phosphorus concentration of 1 mg/l and nitrogen concentration of 20 mg/l, but with an annual average nitrogen concentration of 10 mg/l.

2

Sludge and wastewater treatment

A brief account of sewage treatment and sources of sludge is given here, mainly to clarify the terminology used. A detailed account of biological wastewater treatment processes has been given by Winkler (1981). The somewhat cumbersome term 'wastewater' is used here to describe aqueous waste liquors, rather than the more familiar term 'effluent'. This is because the strict meaning of the term 'effluent' is 'outflow stream'. Domestic or industrial effluent may indeed be an unpleasant, highly polluting aqueous waste liquor, but the effluent from a purification plant should be a clean, healthy liquid. Sewage effluent is thus, strictly, *purified* sewage, although the term is sometimes used confusingly to mean 'effluent on its way to a sewage treatment plant'. In this review, the term 'effluent' will be used as far as possible in its strict sense of 'outflow stream', although it is used ambiguously by many authors.

2.1 SEWAGE

Sewage is a usually dilute aqueous liquid with a content of organic material less than one part in a thousand, made up of a mixture of domestic wastewater, trade and industrial wastewaters, infiltration from subsoil and surface water, mainly from rainfall. The organic content can be gauged from the biologically determined oxygen demand, known as the 'BOD', which is usually about 300 g/m^3 for domestic sewage, equivalent to roughly half a gram per litre of sugar, i.e. half a part per thousand. Although domestic sewage contains unpleasant organic material such as faeces, it is considerably diluted by urine and by washing- and bath-water known collectively as 'sullage'. Indeed, domestic sewage in the USA generally has a lower BOD than that in the UK, which has led to snide comments about the relative amounts of bath-water used in the two countries. Even apparently solid faecal matter is mostly water and contains only 15 to 25% solids. Sewage is still disposed of without treatment, particularly in coastal areas, and the daily discharge of untreated municipal sewage to sea has been estimated as nearly 15 million tonnes in the USA (Gerba & Goyal 1988), and about a tenth of that amount from the UK until it is phased out by the end of the century.

2.1.1 Preliminary treatment

Preliminary treatment of incoming sewage is carried out either as a form of partial treatment or to protect equipment in subsequent treatment processes. It involves removal of heavy solid matter, such as grit, by settling, and of suspended and floating solid matter such as faeces, paper, rags and condoms by **screening**. Screenings are broken up by disintegration devices, then returned to the flow of sewage upstream of the screens to collect inadequately disintegrated material. In this condition, the sewage may pass directly to a biological treatment process, such as a biological oxidation-ditch or circulating-channel system.

2.1.2 Primary sedimentation

More usually, the sewage is subjected to **primary sedimentation**, in which the suspended organic solids settle to give a solids fraction known as **primary sludge**, and a supernatant containing dissolved organic and inorganic nutrients and other solutes as well as residual suspended particles which were too small and/or insufficiently dense to settle out in the primary sedimentation process. The removal of solid material in primary sedimentation may be enhanced by **flocculation**, whereby a flocculent precipitate is formed in the sewage, which collects suspended material on to its flocs as it settles and carries them down into the sludge. A solution containing ions of a polyvalent metal that forms an insoluble hydroxide, such as calcium, ferric iron or aluminium, is mixed into the solution, and alkali is dosed into the mixture to raise the pH level and so precipitate the metal hydroxide. The flocculent hydroxide precipitate settles through the sewage, and as it settles, collects suspended solids and carries them down into the sludge fraction. Flocculation can be enhanced physically by very gentle agitation, and the solid fraction, consisting of the mixture of organic and inorganic solids, is sometimes called **chemical sludge** or **physico-chemical sludge.**

Sludge from subsequent treatment processes, percolator humus or surplus activated sludge, may be returned to the primary sedimentation stage, usually after thickening. The resulting sludge is known as **co-settled primary sludge**.

The terms **sedimentation** and **settling** are used interchangeably. As efficient sedimentation should also effect removal of suspended material which makes the liquid turbid, the process is often called **clarification**, although a liquid could, presumably, undergo sedimentation without necessarily being clarified. Particular care needs to be used in referring to publications by German-speaking authors, as the general term 'purification' is sometimes used with the more restricted meaning of 'clarification' and vice-versa (presumably from *Reinigung*).

2.1.3 Secondary treatment

After primary treatment, the sewage has had a major proportion of the organic material removed in the primary sludge, and is considered to be **partially treated**. It is thus no longer called **raw sewage** and passes to some form of **secondary treatment** or may be discharged to sea as partly-treated sewage. Although the discharge of raw sewage to sea from the UK is to cease before the end of this century, as well as the disposal of sewage sludge to sea, it appears likely that the discharge of settled sewage, counting as (partially) treated sewage, will continue. Discharge of primary effluent to wetlands such as cypress swamps is, for example, carried out in the USA (Scheuerman *et al.* 1986). In the UK, 17% of sewage is currently discharged directly into the sea, and the other 83% is subjected to either primary or secondary treatment at inland treatment works. In the USA, the sewage from 20% of the population receives only primary treatment and 60% receives secondary treatment (Rao *et al.* 1988).

2.1.4 Biological oxidation

Secondary treatment usually consists of some kind of **biological oxidation**, a process in which oxygen, usually from air, is dissolved into the clarified sewage in the presence of a mixed microflora. The micro-organisms then utilize the dissolved oxygen and organic and inorganic substances in the sewage as nutrients, converting them to more organisms and producing carbon dioxide. The utilization of nutrients to form biomass is called **assimilation** and the formation of simple inorganic substances, such as carbon dioxide, from organic substances is called **mineralization**. The breakdown of microbial biomass in the process is called **endogeneous respiration** and organic substances released into the liquid phase from the lysed organisms are utilized as nutrients by the other organisms present. Certain parts of microbial cells are not easily biodegradable, and remain suspended in the liquid as stable, organic solids. The proportions of assimilation, mineralization and endogenous respiration occurring in biological oxidation depend on the design and operation of the process.

2.1.4.1 Stabilization

As the nutrients in the sewage are depleted by the treatment, the sewage becomes less susceptible to putrescence, and is said to be **stabilized**, so that biological oxidation is an example of a **stabilization process**. Stabilization is an important requirement for sewage and sludge treatment, as the treated material is less likely to become malodorous. Stabilization does not necessarily effect the inactivation and/or removal of pathogenic organisms, called **disinfection**, nor does disinfection necessarily effect stabilization. Indeed, re-infection of disinfected sludge is a problem in sludge stabilization. Sewage and sewage sludge should be considered as nutrient biological liquids, like milk, for example, which will undergo putrescence and produce unpleasant odours even after sterilization, if left in contact with the atmosphere.

2.1.5 Suspended-growth systems

There are two main types of biological oxidation processes used as secondary treatments, suspended-growth systems and supported- or attached-growth systems.

The suspended-growth system for biological oxidation is usually known as the **activated sludge process**. In suspended-growth systems, the microbial biomass carrying out the biological oxidation is suspended in the aqueous liquid being treated. Oxygen is dissolved into the suspension by sparging with air, or sometimes high-purity oxygen, and/or by mechanical agitation. Under certain operating conditions, the microbial biomass aggregates to form a flocculent mass which removes some suspended and dissolved materials from the liquid by adsorption on to the flocculent aggregates, and can be separated from the treated liquid by gravity-settling as sludge.

The **activated sludge** is thus a mixture of living, moribund and dead microbial biomass, mostly moribund in practice, with inert organic and inorganic solids. Most of the settled sludge is returned to the aeration stage as an inoculum, but as the biomass accumulates, sludge is periodically removed from the system as waste or **surplus activated sludge**. The liquid in the biological oxidation stage thus comprises a mixture of incoming wastewater, recycled activated sludge and incompletely purified wastewater and is known as **mixed liquor**. The solid matter in the mixed liquor is known as **mixed liquor suspended solids**, usually abbreviated to MLSS. The organic content of the MLSS is taken to be the material lost by heating a sample of the suspended solids and is known as the **mixed liquor volatile suspended solids** or MLVSS. The MLVSS is generally taken as an indication of the amount of biomass in the biological oxidation stage or in the sludge, but it does contain inert organic material, such as cellulose fibres from paper and non-biodegradable residues of dead microbial cells, as well as living and moribund micro-organisms.

2.1.5.1 Operating conditions

It must be emphasized that the condition of the sludge depends on the operating conditions used. The tendency for the mixed microbial biomass to form flocculent aggregates is associated with the organisms being in near-starvation conditions. The principal benefits of floc formation are that the sludge can be separated from the treated wastewater cheaply and efficiently by gravity-settling, and that the flocs remove colloidal and ionic material from the wastewater by adsorption. In addition, the longer a wastewater is subjected to biological oxidation, the lower its residual content of putrescible organic matter, and so the more thoroughly stabilized the wastewater will be. Because most of the sludge separated from the treated wastewater is recirculated to the aeration stage, the average sludge residence time in the system is several times longer than the average hydraulic residence time. In conventional sewage treatment, the hydraulic residence time is about half a day, and the sludge residence time about a week. A secondary treatment process using hydraulic residence times of several days and sludge residence times averaging several weeks is known as **extended aeration**, or sometimes by the more general term **low-rate** treatment. Circulating-channel oxidation-ditch systems, where the comminuted sewage is fed without primary settling directly to the biological oxidation stage, tend to operate as extended aeration systems. A system making particular use of the adsorptive properties of the sludge is known as **contact stabilization**, where the recycled separated sludge is aerated before being brought into contact with the wastewater feed.

When sludge or slurry is stabilized by extended aeration, the process is sometimes called **cold aerobic digestion**.

2.1.6 Attached-growth systems

In attached-growth systems, the microbial biomass effecting the biological oxidation forms a film of slime, sometimes called **bios**, over a solid support called a **packing** assembled in the form of a **bed**. The sewage is distributed over the top of the bed by rotating sprays and percolates through the bed, over the biomass attached to the packing. The packing has a structure sufficiently open for air to pass through freely, owing to wind and/or convection, and oxygen from the air dissolves into the liquid percolating through the packed bed to provide the dissolved oxygen needed for the treatment process. Suspended and dissolved materials are also removed from the liquid by adsorption and attachment to the biomass film. The biomass film gradually accumulates and periodically sloughs off from the packing, is carried out with the treated wastewater and separated by gravity settling in a **humus tank**. The settled slime is known as **humus**, and is the equivalent of the surplus activated sludge from a suspended-growth system. The slime film also supports a population of worms, snails and insect larvae, and *Psychoda*, the 'filter-fly' can be a considerable nuisance to people working or living near these installations.

The treatment unit is confusingly known as a **trickling filter**, **biofilter**, or **percolating filter**, although 'filtration' in the sense of solid–liquid separation has little to do with the purification effect. It is particularly confusing when authors refer to **filters** or **filtration** alone, as filtration, in the chemical engineering sense of separating a solid from a liquid, is commonly used in sewage treatment plants for de-watering sludge. In addition, **sand filtration**, used in drinking-water treatment and **tertiary treatment** of sewage works effluents (section 2.1.8), does utilize both filtration, in the sense of solid–liquid separation, and biological purification using an attached growth of biomass. The solid medium supporting the biomass is conventionally stoneware, but specially fabricated plastic modules are also used. The plastic support systems have the advantages both of being light in weight and of providing a high surface area for the biomass to grow on contained in a small volume of bed.

A recent development in attached-growth systems is to mount lightweight packing material on a shaft so that the packing can be rotated through a bath of sewage. These are known as **rotating biological contactors**, or RBC units.

2.1.7 Stabilization ponds

Stabilization ponds are also known as **lagoons** or, sometimes, **oxidation ponds**, and are the first choice of wastewater treatment system in many parts of the world. Stabilization ponds are cheap to build and operate, but require a large land area. They consist of an area of land bounded by a concrete wall or earthen bank containing wastewater at a shallow depth. Their mode of operation depends on the depth of liquid, the loading of biodegradable organics into the system and the retention time. With correct operation, they can effect secondary treatment to an acceptable standard, and are fed with the supernatant from primary sedimentation. Stabilization ponds are used for domestic, industrial or mixed wastewaters. They can also be fed with treated secondary effluent and used as an additional, or **tertiary**, final treatment, in which case they are known as **maturation ponds** (section 2.1.8).

The large area and shallow depth facilitate access of oxygen from the air to the waste water, and oxygen dissolution can be assisted by circulating or mixing the liquid intermittently. Bacteria and other micro-organisms utilize the dissolved oxygen in aerobic breakdown of dissolved organic nutrients. In sunny climates, algae grow on the liquid surface, utilizing nutrients in the wastewater and carbon dioxide released by bacterial growth, producing oxygen by photosynthesis which is utilized by aerobic bacteria. There can, however, be problems of separating algae from the treated wastewater, and toxin-producing algal strains may also develop. Blooms of toxic algae caused a problem in drinking-water reservoirs in the hot summer of 1990 in the UK, for example.

Typical ponds have an area of one or more hectares, with a liquid depth of a metre or more, and have a retention time of two or three weeks, with intermittent mixing.

A high-rate aerobic pond would have a liquid depth of about half a metre, be intermittently mixed and have a retention time of about a week. An anaerobic or partially anaerobic pond would be two metres or more deep and have a retention time of four or five weeks.

As with other secondary treatment processes, the biomass generated in the pond is removed by settling, which could be in another pond, and/or by tertiary treatment. Sludge also settles onto the floor of the stabilized pond, but only at the rate of several millimetres a year in a well-designed system. Where large quantities of algal biomass are generated, preliminary filtration through a coarse medium must be carried out, to avoid clogging sand filters or micro-strainers in subsequent treatment states.

2.1.8 Tertiary treatment

The treated secondary effluent may be subjected to tertiary treatment, such as sand filtration or micro-straining to remove suspended solids that escaped the secondary settling stage, and/or chlorination to destroy micro-organisms or microbial flocs that escaped sedimentation and traces of refractory organic solutes, and should then be sufficiently purified for discharge into a waterway. Sand filtration involves passing the treated wastewater through a carefully prepared layer of graded sand and has a genuine filtration action in physically separating suspended materials from the liquid phase. The sand grains also develop a microbial film, as in percolator packings, which effect biological nutrient removal. The biological and other material is removed periodically by backwashing, which results in a sludge similar to that produced by percolators, and of course requires treatment and/or disposal in the same way as percolator humus. In the USA, tertiary treatment is also known as 'advanced treatment'. This can cause confusion, as, for example, an 'advanced activated sludge process' is not necessarily a high-technology version of the activated sludge process, but may just be an activated sludge plant used for carrying out tertiary treatment.

There is a growing use of macro-organisms for tertiary treatment, in removing trace nutrients, such as ammonia, nitrate and phosphorus, by assimilation. This can be considered as utilizing eutrophication. Water hyacinths have been used with some success (McAnally & Benefield 1989, Winkler 1984) for this purpose, as have bullrushes. The effluent from secondary treatment processes can also be stored in **maturation ponds**, sometimes called **polishing ponds**, for a few weeks before re-use or discharge to a waterway. Where sufficient land area is available, the additional low-rate biological treatment taking place in maturation ponds can provide a useful back-up in case of a poor performance in a secondary treatment stage (Macdonald & Ernst 1987).

2.1.9 Septic tanks

Septic tanks are used for dealing with the wastes from isolated habitations, and essentially effect primary sedimentation. A commonly used design consists of two chambers in series, the first acting as a sludge sedimentation stage and the second acting as a back-up sedimentation stage to ensure that solid waste is not carried out with the supernatant. The supernatant is taken off for secondary treatment, such as a percolator bed, or discharge into a field at a suitable distance from the habitation, where it can seep into the soil. The settled sludge needs to be removed periodically, every few months, to ensure efficient settling, and is transported either to a local sludge treatment facility or buried. The original purpose of septic tanks was to provide low-rate anaerobic digestion of domestic wastes, but very large tanks are needed to provide sufficiently long retention times. Septic tanks and cesspools are used for treating the sewage from approximately 15% of the population of the USA (Rao *et al.* 1988).

2.2 PHYSICAL AND CHEMICAL SLUDGE TREATMENT

Any purification process separates a material into a purified product and a fraction containing the unwanted material. In wastewater treatment, the sludge is effectively the concentrated pollutant fraction. Organic matter comprises about 70 to 80% of the total sludge solids. The primary sludge contains the biomass and macerated solids from the incoming wastewater, together with any chemicals used in aiding sedimentation. The solids content of primary sludge depends on the nature of the incoming sewage and the operation of the sedimentation system, but is typically about 5% solids. The humus from attached-growth percolator systems usually has a solids content of about 0.75%, but may be as high as 2%. The surplus secondary sludge comprises the residue of the biomass grown and the material adsorbed on to the sludge flocs. The surplus sludge is very dilute, even though it has been through a settling state, and has approximately double the solids content of the mixed liquor in the aeration stage. The operating value of suspended solids concentration in the aeration state, MLSS, is in practice limited by the capacity of the sedimentation state, but is typically 2 to 3 kg/m^3, i.e. less than 0.5%. The surplus sludge thus has a solids content of less than 1% and may be further thickened before being mixed with primary sludge for disposal. The mixture of primary and surplus sludge then contains about 4% solids.

2.2.1 Sludge thickening

Removing sufficient water to increase the solids content of sludge up to about 10% is known as **sludge thickening**. Sludge is effectively a dilute, nutrient biological liquid, so that the immediate aim of sludge treatment is to reduce its volume and to increase its stability as rapidly as possible. In practice, sludge thickening usually involves approximately doubling the solids concentration in the sludge, and so halving the volume for further treatment. The usual method is by gravity-settling aided by gentle agitation, although flotation is sometimes used. Centrifuging may be used where the sludge has poor settling properties. Surplus activated sludge can be concentrated to about 3% solids and primary sludge to about 9% solids. The separated liquid phase is returned to the treatment plant, and the operation must be carried out rapidly before the sludge putresces, generates gas and becomes excessively malodorous. Continuous gravity thickeners have average detention times between four and eight hours. In the UK, about 63% of sludge is subjected to gravity thickening and 11% to mechanical thickening (Newman *et al.* 1989).

2.2.2 Sludge conditioning

As with primary sedimentation, chemicals such as polyvalent metal ions or polyelectrolytes can be added to enhance solid–liquid separation, a procedure known as **sludge conditioning**.

2.2.2.1 Sludge conditioning with polyelectrolytes

Polyelectrolytes are relatively expensive, but are effective in concentrations as low as 0.01% of the sludge. Polyelectrolytes may be cationic, for example polydiallyldimethylammonium salts, anionic, such as polyacrylic acid, or non-ionic, such as polyacrylamide. Polyelectrolytes are the most widely used method of sludge conditioning in both the UK and the EC and are used for 17% of UK sludge and 35% of sludge in the EC as a whole. A further 6% of UK sludge is conditioned using salts of iron, aluminium and/or lime, although 73% of UK sludge, and 39% of EC sludge, is not conditioned at all (Newman *et al.* 1989).

2.2.2.2 Sludge conditioning with lime

Lime has an advantage as a conditioning agent, because it can act as a disinfectant if sufficient lime is added to maintain the sludge pH above 12 for several hours. Furthermore, if quicklime is used, the heat generated by hydration can raise the temperature of the sludge to disinfecting levels. These effects are discussed in more detail further on.

2.2.2.3 Sludge conditioning with acids

Treatment of sludge with acids, such as hydrochloric and sulphuric acids, to produce a pH value of 1.5 has the effect of conditioning the sludge, making de-watering easier, destroying pathogens and solubilizing heavy metals, which can then be removed by ion-exchange (Kiff *et al.* 1984). The procedure is especially effective when used at high temperatures, and is called **hot acid treatment**. Oxidative acid hydrolysis, using hydrochloric acid treatment with an oxidizing agent such as hydrogen peroxide gives selective detoxification and high removal rates of sludge-borne pathogens and parasites.

2.2.3 Sludge de-watering

The term **de-watering** is used with the meaning of reducing the water content of sludge to a level where the sludge has the characteristics of a solid, which occurs when the solids content is about 50%. Processes used for de-watering include filtration, in the sense of solid–liquid separation, centrifuging and drying on sand-beds, whereby water is lost by a combination of drainage and evaporation. The resultant sludge cake is then stored for several months for further biological stabilization to occur. As the process is essentially anaerobic, it tends to give rise to unpleasant odours, particularly in warm weather, although high temperatures generated in the interior of the cake have a disinfecting effect.

About 25% of sludge in the UK is de-watered by filtration, using mostly plate-and-frame presses, about 6% air-dried on beds and 4% centrifuged (Newman *et al.* 1989). The proportions for the EC as a whole are similar, except that centrifuging seems to be more widely used than in the UK, particularly in Spain and Denmark.

2.2.4 Heat treatment of sludge

2.2.4.1 Thermal conditioning

The settling characteristics of sludge can be improved by heat treatments in which the sludge is held at a high temperature and pressure for a suitable length of time. This causes cells in the biomass to break down, including, of course, the cells of pathogenic organisms, thus sterilizing and disinfecting the sludge, as well as improving its settling properties. An example of conditions used in thermal conditioning is treatment at 180 to 200°C and 10 to 20 atm pressure (1000–2000 kPa) for 20 to 30 minutes. Where dissolved oxygen is introduced into the sludge under such conditions, organic matter is oxidized as well as thermally degraded, so the process is known as **wet air oxidation**. Heat treatment is used for conditioning about 8% of sludge in the EC, and is particularly widely used in France and Germany, and to some extent in the Netherlands, but is used for only 1% of UK sludge (Newman *et al.* 1989).

In general, the more severe the conditions, the more the sludge is stabilized, and the lower the organic content of the product. With recent developments, the final product is virtually an ash (Hartmann *et al.* 1989). This special form of wet air oxidation and incineration is capable of achieving 99.9999% destruction of organic material. Wet oxidation converts organic compounds to carbon dioxide and water using dissolved air at high temperature and pressure, about 150 to 250°C and 200 atm (2×10^7 Pa). It does not, however, oxidize ammonia, except for one system that converts nitrogen compounds catalytically to nitrogen gas. In general, the higher the temperature and pressure used, the greater the extent of the oxidation process, providing there is sufficient oxygen present. Subcritical and supercritical water oxidation use less energy than incineration and are carried out in completely enclosed systems, but, like incineration, have a high capital cost. Supercritical water oxidation involves the combustion of organic matter with oxygen in water at temperatures above the critical point of water, 374°C at 22 120 kPa (218 atm). Subcritical water oxidation uses high temperatures and pressures approaching, but less than the critical values, such as 270°C and 150 atm (15 MPa).

Pilot-scale equipment described by Hartmann *et al.* (1989) uses either high-purity oxygen or hydrogen peroxide as the oxidant for a feed of thickened surplus activated sludge containing 5% total solids. The residence time in the reactor is about two minutes at 400°C and 270 atm (27 MPa). The solid residue is an inorganic ash containing 50 to 60% solids, thus effecting 95% removal of total suspended solids, and with 75% of the heavy metal content converted into non-leachable metal oxides in the ash-like residue. The aqueous phase, containing the remaining heavy metals, is apparently returned to the activated sludge plant.

2.2.4.2 Sludge pasteurization

Sludge pasteurization is a relatively simple process in which sludge is heated to a temperature lethal to vegetative microbial cells and is held at that temperature for a time sufficient to have the required destructive effect. The temperature–time requirements specified in the UK Department of Environment guidelines (DoE 1989) range between 30 minutes at 70°C and four hours at 55°C. These values correspond to the relation

$$\text{holding-time (minutes)} = \exp[3.4 - 0.14 \times \{\text{treatment temperature (°C)} - 70\}]$$

It is perhaps the most obvious procedure for inactivating pathogens, and has been widely in use in Switzerland for about ten years. The design and operation of batch and continuous plants for pasteurization of sewage sludge prior to mesophilic digestion has been described by Huber and Mihalyfy (1984). The sludge is pre-heated by heat exchange with heat-treated sludge and maintained at 70°C in tanks using heat from combustion of biogas from the digestion stage. The heat requirements were found to be much the same as that required for (mesophilic) anaerobic digestion, but the electrical energy requirements depended on process design. Biogas produced by anaerobic digestion has been used to pre-heat and pasteurize the digester feed in a submerged combustion system (Kidson & Ray 1984). Residence time in the burner tank was three hours, achieving a temperature between 38° and 55°C, and the sludge was cooled to 35°C before feeding to an anaerobic digester.

2.2.5 Sludge incineration

De-watered sludge is dry enough for disposal by incineration when the solids content is greater than about 30%. At this level, the heat generated by the combustion of the sludge solids is sufficient to evaporate the residual water content and the incineration is said to be **autothermal**. With sludge of higher water content, supplementary fuel has to be used. Incineration, however, leads to problems of atmospheric pollution with carbon dioxide, nitrogen and sulphur oxides and particulate matter carried out with the exhaust gases. Wastewater sludges tend to adsorb heavy metals, which, with microbial pathogens, are the main pollution worry in the land application of sludge. The residual ash creates a disposal problem, as it constitutes 25 to 30% of the original sludge solids, and is dumped on land or at sea, or can be used as a sludge conditioning agent and filter aid in de-watering.

Incineration is unpopular in the UK as it is considered to be expensive, and only about 3% of sludge is incinerated in the UK, although this is likely to increase considerably as dumping of sludge and raw sewage at sea is phased out. For Europe as a whole, the proportion of sludge incinerated is 9%, predicted to increase to 16% in 1994 (Newman *et al.* 1989). Detailed consideration of sludge incineration is beyond the scope of this book but the pollutant emissions from the incineration of waste and pollution arising from contaminated wastewater and ash have been recently discussed by Williams (1990).

2.2.6 Pyrolysis

Pyrolysis consists of destructive distillation of organic material in the absence of air, and gives a char, a liquid fuel, a solution of organic substances and gases, and thus effects sludge disinfection. Recent pilot-scale experiments in Canada (Campbell 1989) gave oil yields of 13% from anaerobically digested sludge and 46% from a mixed raw sludge, with char yields ranging from 40 to 73%. This suggests that the sludge components from which oil is produced by pyrolysis are those that are most readily broken down during digestion. The calorific values of the oil were 32 to 42 MJ/kg and 6 to 23 MJ/kg for the char, with slightly higher values from digested sludge than from raw sludge.

2.2.7 Irradiation

Ionizing radiation can inactivate forms of life such as seeds, microbial pathogens and parasites, and can aid de-watering sewage sludge, settlement of solids and removal of refractory chemicals. The radiation used may be gamma-radiation from cobalt-60, caesium-137 or nuclear-reactor waste products, or high-energy electron beams generated by accelerators. The agents that disinfect sludge in irradiation are fast electrons, whether these originate from the interaction of gamma-rays or in an electron-beam (EB) machine (White 1984). Some of the terminology and units employed were explained by White:

Units: absorbed dose, 1 gray (Gy) = 1 J/kg = 100 rad;

$1\ \text{Mrad} = 10^4\ \text{J/kg} = 2.39\ \text{cal/g}$.

radioactivity, expected number of nuclear transitions from a particular energy state in unit time, 1 becquerel (Bq) = 1/s; 1 curie (Ci) = 3.7×10^{10}/s, so that 1 Bq = 2.7×10^{-11} Ci.

Types of irradiation:

beta-rays: high energy electrons from accelerators ('electron beam machines'). Maximum range approximately 1 mm per 0.5 MeV, maximum intensity at about one-third of maximum penetration range. They are flexible in operation as they can deal with varying loadings by varying beam-current.

gamma-rays: from ^{60}Co (1.17 and 1.33 MeV; half-life 5.27 years; typical installed plant, 45 MCi), or

from ^{137}Cs (0.66 MeV: low energy means self-absorption, so the source must be manufactured in thin form; half-life 30 years, so less frequent re-purchase necessary (but presumably more difficult disposal of spent material).

Equivalence: a 1 MeV accelerator of about 15 kW power-rating is equivalent to 1 MCi of ^{60}Co in terms of energy absorbed.

The inactivation effect of a dose of radiation depends on the water content of the substance irradiated, and temperature and the presence of certain chemical species, notably oxygen, may have a synergistic effect. Recommended doses are 0.5 Mrad for liquid sludge and 1.0 Mrad for dry sludge, which may be appropriate for sludge in horticultural use, but are arguably uneconomic for sludge spread on agricultural land.

Irradiation is fairly widely used in the USA as a method of disinfecting sludge, although it is still relatively uncommon in Europe. Early work on the use of radiation to kill sludge pathogens was carried out by Lowe *et al.* (1956), and the Sandia National Research Laboratories, Albuquerque, New Mexico, has been the centre for sludge irradiation research in the USA.

The Greater Chicago sanitary district has successfully operated a sewage irradiation plant using gamma-radiation (Washington Scientific Trends 1968). A combination of heat and radiation was found to be more effective in destroying *Escherichia coli* and *Ascaris* than either heat or radiation alone (Sivinski 1975).

A recent feasibility study of a process involving pasteurization of de-watered sludge by gamma-irradiation using a cobalt-60 source, followed by rapid composting to produce a soil conditioner, concluded that irradiation can be an effective and reliable sludge disinfection method that leaves the nutrient levels in the sludge unimpaired (Cunningham & Skinwood 1990).

2.3 BIOLOGICAL SLUDGE TREATMENT

An outline of sludge treatments involving biological mechanisms is given here, again partly with the objective of clarifying the terminology in use. The original and principal aim of these treatments was **stabilization** of sludge. However, it was also realized that, with certain operating conditions, several of these processes also **disinfect** the sludge, in other words, destroy or inactivate organisms that cause disease. The disinfection effects are discussed in detail further on, and this section will be confined to providing outlines of the processes currently available and their biological mechanisms.

One of the principal problems with any treatment process is to ensure that every element of the sludge is subjected to the required conditions for the appropriate length of time. The only reliable method of achieving this is with a well-mixed batch process. Inadequate mixing results in non-uniform conditions, so that some elements of the sludge undergo different treatments from the others. It is usually more convenient to operate treatment processes as a continuous or semi-continuous (**fill–and–draw**) procedure. This introduces a risk of untreated or incompletely treated material passing into the output stream, known as **hydraulic short-circuiting**. More technically, continuous operation means that the material passing through the system does not have a single determinable residence time in the treatment process, but has a **mean residence time** with a distribution of residence times about the mean value. In other words, some sludge elements will be subjected to treatments of shorter duration than average, and others to longer durations. A plug-flow system can give continuous operation with batch kinetics, but even then, perfect plug flow is virtually impossible to attain and some elements pass through the system more quickly than others.

2.3.1 Anaerobic digestion

Anaerobic digestion in various versions is the most widely used sludge treatment, used for nearly half of sludge produced in the UK and well over half of EC sludge (Newman *et al.* 1989). Again, these proportions are likely to increase considerably during the 1990s as discharge to sea is phased out and the restrictions on land usage become more severe. Anaerobic digestion is also appropriate for treatment of farm wastes, such as animal slurries, and strong industrial wastewaters containing high concentrations of biodegradable material, for example, those from biological industries such as starch processing and yeast production.

In anaerobic digestion, organic materials are incompletely broken down by microbial action in the absence of oxygen, to give a liquid with a high residual content of organic material. Depending on the conditions used, about 35 to 40% of sludge solids are broken down in the digestion process. The process liberates inorganic nutrients, such as phosphorus- and nitrogen-compounds from the biomass into the liquid phase, and generates a gas, known as **biogas**. The composition of biogas is affected by the digestion conditions, but consists typically of 65 to 70 mole per cent methane and 30 to 35 mole per cent carbon dioxide (methane : carbon dioxide roughly 50 : 50 on a weight basis) with small amounts of other gases such as hydrogen sulphide and hydrogen. On this basis, the gross calorific value of digester gas is about 25 MJ/m^3, and about one cubic metre of methane is generated for each kilogram of organic matter destroyed. Air must be excluded from the system, firstly because oxygen inhibits anaerobic processes, and secondly, because oxygen can form an explosive mixture with methane.

2.3.1.1 Mechanisms of anaerobic digestion

Anaerobic digestion proceeds in several successive stages taking place simultaneously in the digester. In a preliminary **liquefaction phase**, high molecular weight substances, such as polysaccharides and proteins, are broken down to give soluble, low molecular weight compounds, such as sugars and amino acids. In an **acid fermentation phase**, organic nutrients are converted into compounds containing six or fewer carbon atoms and inorganic gases such as hydrogen, carbon dioxide and carbon monoxide. These substances are the substrates for the third **methanogenic phase**, in which methane is generated. The organic compounds involved are lower alcohols and lower fatty acids, which tend to reduce the pH of the digesting mixture. In other processes, sulphates are reduced to sulphide and nitrates to nitrogen gas.

The methanogenic phase is generally regarded to be both very slow, and so rate-determining, and the most sensitive to inhibition. The balance between the production and breakdown of the acids is critical, as the methanogenic organisms are very sensitive to low pH, with a lower pH tolerance limit of about 6.2. If the acid production rate exceeds the acid removal rate, acids accumulate, the pH falls and methane production stops. The accumulation of organic acids also gives the sludge a sour, foul-smelling odour. The slow growth rate of methanogenic organisms and their sensitivity to inhibition means that there is a risk of their being washed out of the digester altogether if the mean sludge residence time is too short, and this has given anaerobic digestion a reputation for instability.

2.3.1.2 Advantages of anaerobic digestion

Anaerobic processes offer several attractive features. The process yields a useful fuel gas, which can be utilized in heat-treatment of sludge. In large sewage works, biogas is, for example, utilized to fuel motors which drive electricity generators or compressors to provide the air supply for aerobic treatments. The cooling-water from these machines then provides low-grade heat for space heating. Materials which are left unaffected or cause problems in aerobic processes, such as cellulose and fatty materials, are broken down in anaerobic conditions, and the hydrogen sulphide formed precipitates heavy metals from solution. In anaerobic processes, the cost of installing and operating aeration equipment is saved. The limitation in aerobic processes is oxygen supply, which is irrelevant in anaerobic processes, so that anaerobic processes can not only deal with high-strength wastes, they actually function better than with low-strength wastes. As anaerobic processes are necessarily enclosed, the smell, aerosol formation and insect nuisances associated with aerobic processes are avoided. Sludge stabilized by well-conducted anaerobic digestion has a pleasant, tar-like odour.

2.3.1.3 Disadvantages of anaerobic digestion

The main problems with anaerobic digestion are its slowness and its sensitivity to inhibition, both of which necessitate long residence times and large-capacity plant. Inhibition is caused by substances commonly found in wastewater sludges, such as heavy metals, anionic detergents and chlorinated hydrocarbons. Methane is a useful fuel gas, but this also makes it an explosion hazard, necessitating carefully enclosed plant and appropriate safety equipment, which adds to the capital outlay. Anaerobic conditions permit sulphate-reducing organisms to form hydrogen sulphide, which is corrosive and extremely poisonous, and needs to be removed from biogas before it is utilized. Some digested sludges have been found to be difficult to de-water by filtration, as they form a compressible filter cake, and centrifuging may be necessary.

2.3.1.4 Types of anaerobic treatment processes

Anaerobic digestion processes for sludge treatment differ principally in their operating temperatures, but there are also variations in the methods of mixing and gas take-off.

The essential elements of a digester are a closed vessel, with provision for the raw sludge feed, digested sludge withdrawal and gas take-off. In the simplest design, there is no agitation, and settled digested sludge is withdrawn from outlets near the bottom of the digester vessel. In more complex designs, the digester itself may be agitated by a mechanical screw pump or by sparging with recirculated biogas, and the mixed sludge is pumped to a second digester vessel for settling. The stabilized sludge is withdrawn from the bottom of the second digester for disposal, and the supernatant returned to earlier treatment stages. In digestion systems for sludge treatment, operation is usually on a **straight-through** basis, and settled sludge solids are not recirculated to the main digester vessel, although sludge recirculation is used in the digestion of strong soluble wastes in the **anaerobic contact process**. With sludge mixing in the digester, pH control systems may be worthwhile, and the state of anaerobiosis may be monitored by measurement of redox potential. The sludge may be heated by heating coils in the digester vessel or by circulating the sludge through an external heat-exchanger. The biogas may be removed for storage in a separate gas-holder, or the digester itself may be fitted with a floating or expandable dome.

The method of charge and discharge is important. The simplest technique is to feed fresh sludge into the digester, and allow the output to overflow into a reception vessel. This involves a risk that a quantity of undigested sludge can flow directly into the output, and can nullify the disinfection effect of digestion. It is safer, but less convenient, to withdraw treated sludge from the system first, then top up the digester with untreated sludge, commonly called **raw sludge**. It should also be borne in mind that digested sludge is still an excellent biological nutrient, and is at considerable risk from re-infection with undesirable organisms such as pathogens. Pathogens finding their way into treated sludge may multiply more rapidly in sterile or pasteurized sludge, because of the lack of competition from the normal robust microbial population of the sludge. It is actually a considerable problem to explain convincingly that something as apparently dirty as treated sludge needs to be handled with almost the same care as milk!

Anaerobic digestion carried out at ambient temperatures, below about 20°C, in temperate climates is known as **cold digestion** and requires retention times as long as 50 or 60 days. In conventional digestion, the sludge is maintained at a temperature between 20°C and 40°C, usually about 35°C, and is known as **mesophilic digestion**. Mesophilic digestion requires retention times of 25 to 30 days, with a minimum of 12 days. In tropical climates, mesophilic digestion occurs at ambient temperatures, but in temperate climates, the heat obtained by burning biogas is usually sufficient to maintain mesophilic digestion. Digestion at temperatures over 40°C is known as **thermophilic digestion**, which requires a retention time of about seven days. There is considerable current interest in thermophilic digestion at temperatures of 55°C and above because of the thermal inactivation of pathogenic organisms effected. In some countries, Switzerland, for example, anaerobic digestion for a protracted period of time is no longer considered satisfactory, because mesophilic anaerobic digestion fails to achieve complete elimination of infectious agents, pathogenic protozoa, bacteria, viruses and parasite eggs present in the original raw sewage (Hamer 1989). Thermophilic aerobic pretreatment provides a possible answer to this objection.

Thermophilic anaerobic digestion with a mean residence time of seven days at 52°C proved to have a superior performance to that of mesophilic digestion with 14 days at 35°C, as well as when both systems were operated with 14-day residence times (Farrell 1984). The thermophilic digester produced about 30% more biogas than the mesophilic digester, but the net gas production was lower, because of the higher heat requirement of the thermophilic system. The odour of the thermophilic sludge caused some concern, however.

A two-stage digestion plant using mesophilic (35°C) and thermophilic (49°C) digestion in series is due to come into operation in New York during 1990 (Farrell 1984). The residence time in each stage is 21 days, and the system has the advantage of giving a sludge solids removal 50% greater than that obtained with conventional mesophilic digestion alone. Farrell (1984) also reported that mesophilic digestion, without preliminary heat treatment, was selected for treatment of waste activated sludge in Los Angeles. This choice was made after extensive comparative pilot-scale trials, in the early 1980s, of mesophilic and thermophilic digestion, with and without preliminary heat treatment.

The length of the retention periods needed shows the importance of thickening sludge before digestion. Digesters with a volume large enough to accommodate ordinary dilute wastewaters for lengthy periods would be impossibly expensive, and even surplus activated sludge is too dilute for digestion without prior thickening.

2.3.2 Aerobic digestion

Aerobic digestion is sometimes called **liquid composting**.

2.3.2.1 Ambient aerobic digestion

Conventional aerobic digestion is effectively the application of the **extended aeration** version of the **activated sludge process** to sludge rather than wastewater. It is usually operated in open tanks in a **straight-through** mode, however, without recirculation.

At ambient temperatures, a hydraulic residence time of two weeks reduces the volatile solids content by roughly half. The process has the advantage of reducing the organic content, expressed as biologically determined oxygen demand (BOD) of both the liquid phase and the suspended solids, to give a stable, easily de-watered sludge. The main disadvantage is the difficulty and cost of supplying sufficient oxygen to a liquid with a high solids content over a protracted period.

2.3.2.2 Thermophilic aerobic digestion

In **thermophilic aerobic digestion**, digestion is carried out in an enclosed reactor at temperatures above 50°C. There is currently considerable interest in this process because the high temperatures used effect sludge disinfection as well as giving fast reaction rates and correspondingly short residence times. A well-designed system with heat conservation can maintain thermophilic temperatures using the heat generated by the oxidation of the sludge solids with very little extraneous energy input.

The development of a full-scale aerobic thermophilic sludge digester (ATSD) of capacity 420 m^3 for processing 5 t sludge dry solids (DS) per day at 60°C has been reported by Morgan and Gunson (1989). As with any sludge stabilization or digestion system, the ATSD process is designed to give odourless, inoffensive, stable, pasteurized sludge with the water volume reduced by evaporation, suitable for disposal at an acceptable cost. The oxidation of sludge solids provides the heat for achieving the process temperature of at least 60°C while removing 40 to 50% of the organic sludge solids. This temperature not only pasteurizes the sludge, but the concomitant faster reaction rates require relatively short retention times to be used in the digester, compared with anaerobic digesters. This implies smaller units and decreased capital costs. The evaporation loss with air operation, estimated as 28%, gives a useful reduction in the volume of treated sludge for disposal, although the cost of supplying the air, as well as the loss of methane produced in anaerobic digestion, must also be taken into account. The nominal retention time was seven days, varying between 5.8 and 8.7 days. This was ample to give thermal inactivation of *Escherichia coli* and *Salmonella* sp., in accordance with the US EPA guidelines of a holding time of at least three days at temperatures above 55°C. The final costs of the sludge thickening and ATSD treatment plant, at 1987 prices, were about £0.5×10^6, with annual operating costs of about £40 000, for a sludge capacity representing a population between 61 000 and 93 000. Problems in operation were however experienced with a sludge heat-exchanger and the odour of the vent gas.

In a study comparing aerobic thermophilic sludge digestion (ATSD) with anaerobic digestion for employment at small sewage treatment works, Bruce (1989) concluded that simple, small-scale ATSD plants will find considerable application for sludge stabilization, following the 1986 EC directive on the use of sewage sludge in agriculture. The thermophilic temperature of 55°C can be maintained, even when the ambient temperature is 0°C, and the design retention time of seven days gave well-stabilized sludge suitable for agricultural use.

2.3.3 Successive aerobic and anaerobic digestion

From the previous sections, it is evident that aerobic and anaerobic digestion complement each other to some extent. Using anaerobic and aerobic thermophilic digestion stages in combination can provide advantages in flexibility of operation unavailable with a single-stage system (Loll 1989). The aerobic-thermophilic stage is used for thermal disinfection and the anaerobic stage provides the energy for maintaining the required temperature.

With the aerobic thermophilic stage upstream of the anaerobic digester, the raw sludge is disinfected and partially stabilized. Partial stabilization is achieved in the aerobic thermophilic digester with a retention time of 12 hours, whereas thorough stabilization requires two to three days. The anaerobic digestion stage then requires a retention time of 10 to 15 days to complete the stabilization.

With the anaerobic digester upstream of the aerobic thermophilic digester, complete stabilization can be achieved in the anaerobic stage, with retention times of 15 or, preferably, 20 days. The aerobic thermophilic digester can then be used for disinfection of all the anaerobically digested effluent or just that part destined for disposal on to agricultural land, as required.

2.3.4 Composting

Composting can be considered as low-rate aerobic digestion, which may be carried out in enclosed reactors or simply in heaps in the open air, known as **windrows**, over a period of several weeks. De-watered sludge is mixed with an inert organic bulking agent, such as straw, sawdust or bark chips, to maintain an open, porous structure that enables moisture to drain out and air to penetrate into the mixture. This prevents the formation of anaerobic regions within the mixture and the creation of unpleasant smells, reduces the moisture content and corrects the nutrient carbon-to-nitrogen ratio. The oxidation of the organic material can generate temperatures as high as 55°C, and if this is maintained for over four hours can effectively pasteurize that portion of the compost. The compost needs to be mixed to ensure that all the sludge is subjected to the required treatment, and while this can be effected by mechanical agitation in reactor systems, heaps need to be turned regularly, weekly, for example, as in winter, pasteurizing temperatures may be attained only in the centre of the heaps. The size of the heaps is important, as small heaps lose heat rapidly and may not reach pasteurizing temperatures, while in large heaps, access of oxygen to the centre of the heap may be restricted, with a concomitant production of anaerobic conditions. Composting can be accelerated by additional forced aeration from underneath the heaps, as in the 'Beltsville' aerated pile method.

The product makes a useful soil conditioner, and there is considerable current interest in this material for that purpose. At the moment, gardeners and professional horticulturists use large quantities of sterilized peat for soil conditioning, and there is now concern amongst conservationists about the destruction of peat bogs to meet this demand. Composted sludge may not only provide an alternative and income-generating means of disposing of sludge, but may also prevent damage to the environment. Composting also provides a timely means of utilizing waste straw. Currently, ten million tons of straw are burnt in the UK each year, but stubble burning will be banned after 1992.

Composting in enclosed reactors has many similarities to thermophilic aerobic digestion, for which the term **liquid composting** is sometimes used. The principal difference is that composting is used for de-watered sludge, and generally has a bulking agent mixed with it.

2.3.5 Lagooning

This is probably the simplest method of sludge treatment, and involves storing the sludge in shallow reservoirs held in by earth banks for over a year, and at least three months. This is sometimes called **cold digestion**. It has the advantage of simplicity, but requires a relatively large land area and the biogas generated is lost to the atmosphere. In view of current concern about the 'greenhouse effect', it should be noted that methane is a very much more effective absorber of infra-red radiation than carbon dioxide. In **accelerated cold digestion**, digestion is accelerated by the addition of mesophilically digested sludge.

2.4 EFFECTIVE SLUDGE TREATMENTS

Examples of sludge treatments given in the 1989 Code of Practice considered to be effective are as follows, with effectiveness presumed to be in terms of pathogen reduction rather than the less easily quantifiable odour nuisance.

Mesophilic anaerobic digestion should involve a two-stage treatment, the primary stage having a mean retention period of either at least 12 days at a temperature between 32°C and 36°C, or of at least 20 days at a temperature between 22°C and 28°C, followed by a secondary stage with a mean retention period of at least 14 days.

Sludge pasteurization should involve temperature–time conditions corresponding to a minimum of four hours at 55°C or 30 minutes at 70°C, *and* be followed *in all cases* by primary mesophilic anaerobic digestion.

Thermophilic aerobic digestion should have a mean retention period of seven days, with all sludge being subjected to a temperature of at least 55°C for a minimum of four hours.

Composting in windrows or aerated piles should ensure that the compost is maintained at a temperature of 40°C for at least five days, and during this period, the temperature within the body of the pile should attain 55°C for at least four hours. This treatment must then be followed by a maturing period long enough to ensure that the composting reaction is complete, previously specified as two months.

Lime conditioning of liquid sludge should ensure that for at least two hours the pH is not less than 12.0, after which the sludge may be used directly.

De-watering of untreated sludge after conditioning with lime or other coagulants should be followed by storage of the de-watered cake for at least three months. The cake from de-watered sludge previously subjected to primary mesophilic anaerobic digestion should be stored for at least 14 days.

Storage of untreated liquid sludge should be for at least three months.

2.5 FARM WASTES

Most animal wastes and silage effluent from livestock farms have to be returned to the land, ideally as a replacement for fertilizers so that inorganic nutrients are continually recycled (Owen & Stansfield 1989). They thus constitute a potential source of water pollution, although some may be used as an energy source or sold after drying or composting. To minimize land run-off, winter-produced slurry should be stored for application in the spring and summer. This requires soundly designed and constructed storage facilities, which can be expensive. Good waste management can be summarized as avoiding excessive application, based on knowledge of the nutrient content of the material, the soil type and the level of the water-table.

Land disposal problems also arise from the resulting smell, and the limitations on the amount that can be spread in any given length of time. The smell problem can be alleviated by stabilizing the slurry before spreading onto land or by injecting slurry into the soil. Injection techniques are currently undergoing active development.

2.5.1 Farm waste management

The management of livestock manure has been reviewed by Bailey (1990). **Manure** is the general term for animal excreta, and may include animal bedding material, wash-water and rain as well as faeces and urine. The term **slurry** describes manure which is sufficiently fluid to pump or flow under gravity, and **solid manure** is manure that is sufficiently solid to be stacked in heaps. Cattle slurry then contains about 10 to 15% solids, solid manure from cattle, with straw bedding, contains about 20 to 25% solids. Actual values of manure composition depend on the type of animal and how it is kept. For example, broiler chickens have wood-shavings bedding, while battery hens for egg production have no bedding, and pigs fed on swill produce three times as much excreta as pigs fed on dry meal. Pigs and poultry are generally kept in permanent housing, while cattle tend to be kept housed during the winter months. This is reflected in manure production and disposal, as pig and poultry manure is produced throughout the year, while cattle manure is deposited on land during grazing in the summer months. Spreading on land is the disposal method used for nearly all manure, and consents for discharges to sewers and waterways are rare.

Solid manure and slurry are collected and stored to await a convenient disposal opportunity. Solid manure is stacked in heaps, and storage gives time for breakdown of the bedding material in what is effectively aerobic composting. The centre of the stack may reach temperatures as high as 60°C, which will inactivate parasite ova and cysts. Care must be taken that run-off from the stacks does not contaminate water sources. Battery-hen manure may be air-dried to increase its solids content. Slurry is stored in tanks, or lagoons with earth banks or wooden or concrete walls. During storage, the slurry undergoes anaerobic digestion. Before storage, solids can be mechanically separated from the slurry and stacked, while the liquid fraction is stored. Slurry stores and farmyard water are involved in most reported pollution incidents from farm waste, while relatively few incidents involve run-off from land spread with manure. The proportion of land run-off incidents increased considerably during a dry year, which was attributed to cracked soils. Silage effluent from silage stores, anoxically fermented grass, is another major cause of farm pollution.

2.5.2 Rate of land application

The annual requirement of nitrogen applied to a hectare of land as fertilizer is about 200 to 400 kg nitrogen, which would be supplied, for example, by the manure from three or four head of cattle or 30 or 40 pigs. The development of intensive animal husbandry has given rise to waste disposal problems in two ways. The intensification means not only that large quantities of waste are generated, but it is also generated in concentrated areas at some distance from the land where it could be utilized. Indeed, in small, densely populated areas, there may just be insufficient land available at all to take up the animal waste generated.

The quantities and compositions of animal wastes vary not only with the type of animal, as would be expected, but also according to the way the animals are kept, the breed of animal and the purpose for which it is reared. The volume of slurry generated and its solids content, for example, depend on how the waste is collected and how much water is used to flush out the animal pens. Approximate generation rates of animal wastes have been collated from various reports (Winkler 1984).

Table 2.1

		Cattle	Pigs	Poultry
Wet manure (kg/head-day)	Range	17 to 57	1.3 to 7.9	0.05 to 0.18
	Typical value	40	3.5	0.11
Total solids in slurry (%)	Range	12 to 27	10 to 17	10 to 50
	Typical value	13	12	28
Total solids (kg/head-day)	Range	1 to 6	0.23 to 0.73	0.011 to 0.052
	Typical value	4	0.5	0.03

2.5.3 Slurry stabilization

Anaerobic digestion appears to be the most appropriate method of treatment for animal slurry, as not only does it stabilize the slurry, but the surplus gas can be used for driving electricity generators and space heating of animal accommodation. It is essential to maintain a high concentration of digestible solids in the slurry, with a recommended minimum of 6% (Chesshire 1986), so that a minimal amount of water should be used in flushing out the slurry during collection. Animal wastes are also collected in the form of farmyard manure, in which litter, such as straw, is used for absorbing moisture from the animal waste. This is traditionally treated by stacking to allow uncontrolled composting. If farmyard manure is subjected to anaerobic digestion, very long retention times of about 45 days are needed to break down the fibrous litter material (Chesshire 1986). Accounts of digestion systems for animal slurry stabilization and biogas generation have been given by Chesshire (1986) and Loll (1986).

3

Helminth parasites in sludge

The term **helminth** is from the Greek for 'worm', and there are thousands of species of parasitic helminths and even more that are free-living. A parasite can be defined as an organism that obtains its nutrients from the tissues or body fluids of another organism, called the **host** (Snowdon *et al.* 1989a). In terms of population biology, helminth parasites have been defined as metazoan macroparasites, slower growing than microparasites such as viruses, bacteria and protozoa, with longer generation times, and with low-rate or no direct multiplication within the host. As the immune response of the host to helminth infections is transient and also depends on the number of parasites present, macroparasitic infections tend to be persistent with continual re-infection of the host (Anderson & May 1979 cited by Warren 1988). The helminths include several worms which are important as parasites of man, livestock and crops, and may be transmitted through human and animal waste products. The helminth parasites considered here are those important in sewage and sludge treatment, which inhabit the intestines of humans, and in some cases animals.

Helminths are considered in two groups, flatworms and roundworms. Roundworms are known as **nematodes** and flatworms include tapeworms, called **cestodes**, and flukes, known as **trematodes**. Helminths exist in at least two forms. The first is an actively growing form inside the host which produces eggs or ova. The ova pass out of the host in the faeces and constitute, or develop into, a second form, which is both resistant to adverse conditions and the means of infecting a new host and establishing a new growth. The infective form needs to be resistant in order to survive the acid conditions in the host's stomach in passing through to the intestines.

The occurrence and survival of helminth parasites generally will first be considered briefly, followed by a consideration of the effects of sludge treatments on helminths generally. More detailed consideration of selected key species of flatworms, principally tapeworms, and of roundworms, their hatching and the general effects of wastewater and sludge treatments on them is given in the following two chapters.

3.1 OCCURRENCE OF HELMINTH PARASITES

Schultz (1974) drew attention to the existence of parasitic disease in the well-developed westernized countries, when he found a high proportion of the population of South Carolina was infected with *Ascaris* and *Trichuris*. A recent survey of sludge quality and treatment in 24 selected sewage treatment plants in Germany (Strauch 1989) showed that unidentified parasite ova were found in over 70% of samples of raw sludge in winter and in 50% in summer. In data cited by Snowdon *et al.* (1989a), samples of raw and treated sewage contained 0.78 and 0.46 helminth eggs per litre respectively in one study (Wilkens 1982), and samples of mixed human and farm waste were found to contain *Ascaris, Diphyllobothrium, Enterobius, Taenia* and *Trichuris* (Arkhipova 1979). A survey of sludge samples from 27 US sewage-treatment plants (Reimers *et al.* 1982) found as the principal species, in order of incidence, *Ascaris, Toxocara, Trichuris, Hymenolepis diminuta* and *Capillaria* eggs, *Entamoeba coli* and *Giardia* cysts, eggs of *Trichosomoides* and *Ascaridia*, and *Coccidia* oocysts. Another survey of parasites in wastewater listed 46 pathogenic protozoa and helminths, although enumeration of parasite species may be distorted by the difficulty of recovering protozoan species from samples, and a proportion of eggs and cysts recovered may not be infective. Examples of the incidence of parasite infection in the USA cited by Snowdon *et al.* (1989a) are that in humans, the prevalence of *Ascaris* is about 1%, incidence of *Taenia saginata* infection is less than 1%, and of specimens submitted to state laboratories in 1977 and 1978, 0.05% showed evidence of taeniid tapeworm infection.

Analysis of 145 samples of different types of sludge from a wastewater treatment plant in southern France, some after five years' storage, showed that over 87% contained eggs of either nematodes or cestodes (Schwartzbrod *et al.* 1987). Nematode eggs predominated, principally those of *Ascaris* and *Trichuris*, with some of *Toxocara*. Of the cestodes, *Hymenolepis* was slightly more common than the nematode *Toxocara*, and somewhat surprisingly, *Taenia* eggs formed only about 5% of the eggs found. A free-living non-parasitic Ancylostomides nematode was not counted, although it was present in many of the samples.

3.2 PERSISTENCE OF HELMINTH PARASITES

The survival of parasites outside the host's gastro-intestinal tract depends on the environmental conditions. Death of *Taenia saginata* is reported to occur within five minutes at 71°C and of *Necator americanus* 50 minutes at 45°C (Gotaas 1953).

In suitable environmental conditions, the eggs of *Ascaris, Trichuris* and *Toxocara* can survive in soil for several years, and, for example, the eggs of *Ascaris* have been found to remain infectious even after five to seven years in soil (Müller 1953 cited by Snowdon *et al.* 1989a). The eggs of *Ascaris suum* survive from one day to one year, depending on temperature and water activity, but on leafy vegetables, were found to be completely inactivated after 27 to 35 days (Bürger 1982, cited by Snowdon *et al.* 1989a).

The eggs of the nematodes *Ascaris, Trichuris*, and, to some extent, *Toxocara* were found to survive in stored sludge for over three years (Schwartzbrod *et al.* 1987). In the survey of 24 German sewage treatment works mentioned earlier (Strauch 1989), sludge treated by conventional aerobic or anaerobic digestion showed unidentified parasite eggs in over 45% of samples in winter and 54% in summer, compared with 71% and 50% of raw sludge samples, although the infectivity of the eggs was not evaluated.

The survival of helminth eggs on crops and soil has been summarized in surveys of published data (Stien & Schwartzbrod 1990, Feachem *et al.* 1983). *Taenia* eggs survived on soil for a period of 160 days during spring and winter, and 180 to 210 days in autumn and winter. On grass, *Taenia* eggs survived for 30 days in winter and 57 days in late summer, while on hay (dried grass), *Taenia* eggs have been reported to survive in winter for 22 days and as long as 210 days. *Taenia* eggs survived 80 days at 10°C in silage, anoxically fermented grass. The survival of *Taenia* eggs on soil was 159 to 180 days during spring and summer. *Ascaris* eggs have been reported as surviving for two to seven years on soil and irrigated soil and about a month on plants, fruit and vegetables. This makes an interesting comparison, as *Ascaris* ova survive much longer on soil than those of *Taenia*, but *Taenia* appear to survive better on plants. Infections by *Ascaris* and nematodes generally were reported in a survey of epidemiological studies of faecal samples from people in contact with wastewater. The WHO guidelines on microbiological quality for the use of wastewater in agriculture recommend a level of less than one intestinal nematode egg per litre, of *Ascaris* spp., *Trichuris* spp. or hookworms (Stien & Schwartzbrod 1990). This should be achievable by treating the wastewater in a series of stabilization ponds with a sufficiently long retention time.

Birds are thought to contribute a secondary role in transmission of parasite ova by acting as vectors between the sewage treatment works and pastures and water reservoirs (Silverman & Griffiths 1955, Crewe 1977, 1983). An important point made by Snowdon *et al.* (1989a) is that the ability of man to travel increases the transmission of parasites in human wastes. Disposal of human-derived wastes in the environment increases the likelihood of introducing a new parasite to a previously uninfected region where natural resistance factors have not developed.

3.3 HELMINTH SURVIVAL IN SEWAGE AND SLUDGE TREATMENTS

This section gives a brief account of the effects of sewage and sludge treatments on helminths generally. Where the effects of treatments on particular helminth species have been investigated, however, they are discussed under those species.

The increased prevalence and incidence of parasitic disease in Europe after the First World War stimulated Cram (1926) to investigate the effect of wastewater treatment processes on the survival of helminth ova and protozoon cysts. The likelihood of military personnel returning from the Second World War after spending time in areas where parasites were prevalent led to a further study by Cram (1943) which revealed that after $2\frac{1}{2}$ hours of primary settling, *Ascaris* and hookworm ova settled out of the effluent into the settled primary sludge, while cysts of *Entamoeba histolytica* remained in the effluent. Evidence in Cram's report also indicate that both amoebic cysts and helminth ova may pass through attached-growth percolator systems and survive activated sludge treatment.

3.3.1 Parasites in sewage treatment

The association between cysticercosis in cattle and the land application of sewage sludge has been long established (Roberts 1935). The effective removal of *Taenia* eggs at wastewater treatment plants would reduce the risk of animals exposed to sludge and wastewater. A one-year survey of five sewage plants near Johannesburg was carried out by Hamlin (1946) who found *Taenia saginata* ova present in the incoming sewage, settled sewage, primary filter effluent, humus tank effluent and activated sludge effluent. The same study suggested that *Taenia* eggs did not pass secondary sand filters or underdrained land filtration areas, and the use of sand filtration to remove eggs prior to irrigation was recommended.

An investigation by Wright *et al.* (1942) on the effects of sewage treatment processes on ova and cysts of intestinal parasites found helminth eggs, *Ascaris, Trichuris* and *Hymenolepis*, in samples collected from all stages of wastewater treatment. However, *Entamoeba histolytica* was not isolated from any of the samples, although a similar species was found in the digestion tanks. The existence of helminth ova in sludge samples has been regularly reported (Bhaskaran *et al.* 1956, Krige 1964, Watson *et al.* 1983, for example).

3.3.2 Parasites and primary sedimentation

The most efficient process for the removal of *Taenia* and other parasite ova is primary sedimentation (Newton *et al.* 1949, Bhaskran *et al.* 1956, Forstner 1970a, Carrington 1980a). The time-period for effective removal has been observed to be two hours (Table 3.1).

Table 3.1 — Sedimentation of *Taenia saginata* eggs in a 475 mm column of raw sludge (adapted from Newton *et al.* 1949)

Number of tests	Settling period	Percentage removal of eggs
9	15 minutes	51
6	30	65
3	60	81
2	120	98

As *Taenia saginata* eggs have a diameter of about 40 μm and a specific gravity of 1.3, their Stoke's Law velocity of settling in water at 15°C has been calculated as 0.83 m/h, compared with experimental values of about 1 m/h, and with a diameter of about 50 μm and a specific gravity of 1.111, the Stoke's Law settling velocity of *Ascaris lumbricoides* ova in water at 15°C has been calculated as 0.48 m/h (Pike 1990). The effective liquid upward flow velocity is conventionally between 0.5 and 1.5 m/h in primary and secondary settling tanks, so that ova of *Taenia saginata* and *Ascaris lumbricoides* are unlikely to settle out in primary sewage treatment, and residual ova in secondary clarification, without the eggs aggregating or attaching to larger and/or heavier solid particles. Silverman (1955) acknowledged that tapeworm eggs settled 457 mm in two hours, but noted that most primary sedimentation tanks were $1\frac{1}{2}$ to 4 m deep and were subject to turbulence from the constant inflow of sewage. Silverman doubted whether sedimentation was capable of removing a high percentage of tapeworm eggs from primary effluent.

Slaked lime is often used as a sludge conditioner before filtration (dewatering). This raises the pH to between 9 and 13, which rapidly destroys vegetative bacterial cells. Viruses are destroyed by the release of free ammonia at pH levels of about 12, but parasite eggs appear to be insensitive to high pH levels. When quicklime is used, temperatures of 70 to 80°C are attained, which rapidly kills pathogens. The main difficulty in lime treatment is the incomplete penetration of lime into sludge aggregates, which is much less of a problem with the heat effect.

3.3.3 Parasites in aerobic treatment

Panicker and Krishnamoorthi (1981) investigated the parasite ova and cyst reduction in oxidation ditches and aerated lagoons. They reached several conclusions. First, parasites were reduced by 90 to 100% in an oxidation ditch operating at its optimum level, and most of the cysts and eggs were removed by settling, with the protozoal cysts settling slowly. Lagooning removed 75 to 100% of parasites, and so the removal efficiency in the aerated lagoon was less than that in the oxidation ditch. It was, however, suggested that the removal levels would have improved had effluent from the lagoon been allowed to settle in a settling tank (Table 3.2).

Table 3.2 — Removal of parasites in aerated lagoon and oxidation ditch (adapted from Panicker & Krishnamoorthi 1981)

Parasite	Raw wastewater: maximum number of parasites per litre	Effluent from oxidation ditch		Effluent from aerated lagoon	
		maximum number per litre	minimum percentage decrease	maximum number per litre	minimum percentage decrease
Entamoeba histolytica	600	120	70	96	67
Giardia lamblia	184	30	60	30	65
Ascaris lumbricoides	120	4	84	20	70
Hymenolepis nana	46	7	72	20	48
Hookworm	80	14	80	40	50

All species showed a *maximum* decrease of 100% in both treatments.

Arther *et al.* (1981) isolated parasite ova from digested and lagooned sludge from Chicago, USA. The lagooned sludge had previously been anaerobically digested. Although none of the isolated ova, *Ascaris* spp., *Toxocara* spp., *Toxascaris leonina* and *Trichuris* spp., embryonated after isolation, further development did occur when the eggs were exposed to favourable conditions. Total embryonation was 55.45% for ova recovered from fresh anaerobic sludge and 17.0% for those recovered from the lagooned sludge (Tables 3.3 and 3.4).

3.3.4 Parasites and sludge conditions

The effects of sludge treatment on helminth parasites were demonstrated by the analysis of the 145 samples of different types of sludge from a wastewater treatment plant in southern France (Schwartzbrod *et al.* 1987) (section 3.1.1), some after five years' storage. Chemical conditioning, with 20 to 25% lime plus 3.5 to 6% ferric

Table 3.3 — Embryonation of parasitic nematode ova from anaerobically digested wastewater sludge (adapted from Arther *et al.* 1981)

	Number observed (total of 4 samples)	Number of ova embryonated
Ascaris spp.	33	21
Toxocara spp.	64	34
Toxascaris leonina	8	5
Trichuris spp.	5	1
		Overall: 55.45% embryonation

Table 3.4 — Embryonation of parasitic nematode ova from lagooned sludge (adapted from Arther *et al.* 1981)

	Number observed (total of 4 samples)	Number of ova embryonated
Ascaris spp.	37	9
Toxocara spp.	39	4
Toxascaris leonina	7	3
Trichuris spp.	17	1
		Overall: 17% embryonation

chloride, reduced the recovery of eggs, but mesophilic anaerobic digestion, at 35°C with a 15- to 20-day retention period, had little effect on the numbers of eggs. The types of sludge tested and their egg contents were activated sludge (125 eggs/kg), digested sludge (105 eggs/kg), conditioned sludge (115 eggs/kg) and de-watered sludge (83 eggs/kg). Most of the eggs were those of *Ascaris*. *Trichuris*, the next most common, appeared to be the most resistant to mesophilic anaerobic digestion, but was susceptible to the chemical conditioning. *Toxocara* and *Hymenolepis* appeared to be little affected by sludge treatments. Surprisingly, only a few *Taenia* eggs were found, which were apparently unaffected by mesophilic anaerobic digestion, but were removed completely by chemical conditioning and de-watering.

The effects of the type of sludge, its temperature and its acid content on the survival of different helminth ova were investigated by Kiff and Lewis-Jones (1984). To investigate the effects of temperature, eggs of *Ascaris suum, Fasciola hepatica* and *Monieza* were incubated in primary and digested sludges at 45°C and 60°C and in distilled water at 30°C, 55°C and 60°C, for periods of time ranging from two minutes to 24 hours, then removed and incubated in water at room temperature for several weeks.

Both sludges gave complete inhibition of *Ascaris suum* egg development at 45°C and 60°C and also at room temperature. In water, the proportion of damaged *Ascaris suum* eggs increased with increasing temperature and exposure time. With *Monieza*, damage to eggs was observed at all temperatures above 50°C, even with only a few minutes' exposure. In both primary and digested sludges, all eggs were damaged by exposure for 24 hours at 45°C and two minutes at 60°C. Total inhibition of normal development of *Fasciola hepatica* eggs resulted from exposure at all temperatures of 50°C and above, and the proportion of eggs damaged was directly proportional to the exposure time at that temperature.

The effects of pH were investigated by incubating parasite eggs in phosphate buffers at a range of pH levels at room temperature, 27°C and 37°C. Acid pH levels inhibited normal development of *Ascaris suum* eggs at all temperatures, but highly alkaline buffers allowed development to the infective larval stage.

To test the effects of different organic acids, parasite eggs were incubated in solutions of organic acids at concentrations approximating those in sewage sludges, viz. acetic acid 400 mg/l, propionic acid 300 mg/l, isobutyric acid 40 mg/l, *n*-butyric acid 150 mg/l and isovaleric acid 40 mg/l, at room temperature and 27°C. Organic acids produced almost complete inhibition of development of *Ascaris suum* eggs at both temperatures. *Monieza* eggs were damaged, with all the organic acids having much the same effect, but the exposure time was important. None of the organic acids prevented hatching of *Fasciola hepatica* eggs at either temperature.

3.3.5 Parasites and sludge digestion

The effectiveness of different sludge disinfection procedures was checked by microbiological examination of disinfected sludges from eight sewage-treatment plants in Germany using different disinfection procedures, pasteurization, aerobic thermophilic stabilisation (ATS), ATS followed by mesophilic anaerobic digestion, treatment with slaked lime or quicklime, and composting in windrows and in closed reactors (Strauch 1989).

Parasite ova were found in winter in 70.8% of raw sludge samples and in 45.8% of treated samples; in summer, in 50% of raw sludge samples and in 54.2% of treated sludge samples. The infectivity of the eggs could not be evaluated. In the aerobic thermophilic processes, *Ascaris* eggs were not detectable after four hours at 50°C, 30 min at 55°C and 15 min at 60°C. These tests showed that, of the eight sewage-treatment plants, two of them (aerobic thermophilic digestion, liming), had treated sludge samples positive for parasites.

Black *et al.* (1982) studied the effect of mesophilic anaerobic and aerobic digestion of sludge on the survival rates of the nematode ova *Ascaris suum, Toxocara canis, Trichuris vulpis* and *Trichuris suis*. They referred to the current practice in wastewater treatment of following primary settling with an activated sludge system and anaerobic or aerobic digestion. In addition they acknowledge that these three systems combined to remove some parasites such as *Eimeria, Trichinella, Toxoplasma, Entamoeba* and possibly *Giardia* and *Schistosoma*, a viewpoint supported by others (Fox & Fitzgerald 1976, Fitzgerald & Prakasam 1978). However, parasites such as *Ascaris, Toxocara, Trichuris* and *Toxascaris* are known to survive digestion (Fitzgerald & Ashley 1977, Theis *et al.* 1978, Arther *et al.* 1981, Kabrick & Jewell 1982). Black *et al.* observed that over a 15-day retention period, 23% of *Ascaris* eggs were destroyed by anaerobic and 38% by aerobic digestion, while 11% of *Trichuris* ova were destroyed in an aerobic digester. *Trichuris* eggs in an anaerobic digester and *Toxocara* in either anaerobic or aerobic digesters were found not to be destroyed.

Newton *et al.* (1949) noted the efficiency of sand filtration in removing eggs and concluded that it was a suitable method for preventing the spread of *Taenia* eggs in primary effluent. They also found that anaerobic digestion at ambient temperatures was a poor destroyer of eggs, with normal ova recovered from sludge after six months' digestion at 24°C to 29°C, attached-growth percolators were observed to remove 62 to 70% of *Taenia* eggs, while five months of anaerobic digestion had little or no effect on *Taenia* eggs.

The likelihood that slaughterhouse and restaurant wastes contained larvae of *Trichinella spiralis* was investigated by Fitzgerald and Prakasam (1978), who found that encysted larvae of *T. spiralis* survived a maximum of 96 hours in a batch digester. Fitzgerald (1982) tested for transmission of ascarid helminths from anaerobically digested sludge, and his experiments involved the exposure of pathogen-free pigs to anaerobically digested municipal sludge. Sixty out of 215 pigs were subsequently found to have ascarids, indicating that the sludge contained viable ova. The same study showed that sludge previously lagooned for over two years also contained viable ascarid ova.

3.3.6 Parasites in aerobic digestion

The high temperatures reached during aerobic digestion produced a greater reduction in viable parasite ova than that obtained with anaerobic digestion (Kabrick & Jewell 1982). Using a laboratory digester, Pike *et al.* (1983) investigated the effect of 10 days' digester retention at 35°C and found that only 6% of ova were recovered from the digester at 49°C, while 12% were recovered from the digester at 35°C, although in general the numbers of ova recovered were considerably smaller than expected. Only 1% of the ova recovered from the digester at 49°C developed into larvae.

The survival of helminth eggs during the windrow composting of activated sludge and chemical sludge from wastewater treatment plants with garbage was investigated (Thévenot *et al.* 1985). After 120 days' composting, helminth eggs were present in 60% of samples. Nematodes, *Ascaris, Trichuris,* and *Toxocara*, and the cestode *Hymenolepis* were recovered, but no trematode eggs were recovered.

3.3.7 Irradiation

In previous work cited on sewage sludge utilization Waite *et al.* (1989) found that doses of gamma-radiation of about 400 krad destroy eggs of *Ascaris* spp. in sewage sludge. Viruses were reduced by much less an extent than bacteria or parasites, indicating that the inactivation efficiency of organisms by electron irradiation is a function of the size of the organism, as with other forms of radiation. With high-energy electrons, the larger the target, the more effective the disinfection. A laboratory evaluation of the effects of different electron doses on selected micro-organisms showed that eggs of *Ascaris suum* and *Schistosoma mansoni* were inactivated by very low doses of high-energy electrons (Levaillant & Gallien 1979, cited by Waite *et al.* 1989). A dose of only 70 krad was sufficient for complete elimination of the *Schistosoma* trematode. Electron-beam radiation in doses of 480 krad for 24 seconds destroyed 97% of the ova of *Ascaris suum, Trichuris suis, Fasciola hepatica, Capillaria obsignata*, although oocysts of *Eimeria tenella* and third stage larvae of *Oesophagostomum* survived (Enigk *et al.* 1975).

3.4 PLANT PARASITES

Most of the parasites considered here are important as pathogens of humans and animals. However, if sewage sludge used in agriculture contained plant pathogens, the effect on crops could cause serious economic problems. Plant pathogens could be dispersed, for example, in sludge derived from sewage containing vegetable processing wastes that had been inadequately treated.

3.4.1 Potato cyst nematodes (PCN)

The plant parasites of most immediate concern are the potato cyst nematodes (PCN), *Globodera rostochiensis* or *Heterodera rostochiensis*, the yellow potato cyst nematode, and *G. pallida* or *H. pallida*, the white potato cyst nematode. *Glodobera* and *Heterodera* are synonyms for the same genus: to avoid errors, the name used in this review will be that used by the original author. The root systems of plants are damaged by eelworms so that the plants are unable to obtain essential water and nutrients (Linfield 1977). Potato cyst eelworms are one of the greatest agricultural pests in the world, causing economic loss of £33 million in 1977 (Stone 1977). This clearly imposes a limitation on the use of sewage in horticulture and agriculture, as a result of the *possibility* that sludge might contain plant parasites.

A grower intending to produce seed potatoes must first obtain permission, which is granted only after the soil is tested and declared free from PCN. Land in which viable cysts of PCN are detected must not be used for seed production for six to 12 years, under the Potato Cyst Eelworm (Great Britain) Order 1973 (Bruce *et al.* 1990) and EC regulations prohibit the export of bulbs and seed potatoes containing parasites. A viable cyst contains several hundred eggs, typically 300 (McCormack & Spaull 1987). Soil sampling for cysts and eggs, the normal method for detection, can show that cyst nematodes are present, but cannot guarantee that they are absent, since detection is difficult below a population of 25×10^5 cysts/ha (Linfield 1977). The presence of dead empty cysts in soil is also taken as evidence of eelworm infestation, so that sludge treated to inactivate cysts, but containing dead cysts, is unacceptable for use in seed potato growing. Clearly, the introduction of PCN on to previously infested land would have serious consequences for potato growers lasting for many years afterwards.

The PCN are endoparasitic nematodes less than 1 mm in length which are found in field soils. A number of other species exist which are capable of attacking a wide range of root crops (Table 3.5).

3.4.2 PCN in sewage sludge

PCN are not found widely in sewage sludge, but are likely to be present in numbers below the detection limit, and it is likely that cysts from washing and processing of potatoes may be present in wastewater and sludges. The presence of cysts in some sludges has been demonstrated in tests by the Anglian Water Authority at a level

Table 3.5 — The most common species of *Heterodera* (adapted from Linfield 1977)

Species	Common name	Host crops
H. rostochiensis	Potato cyst eelworm	Potato, tomato, aubergine
H. pallida	Pale potato cyst nematode	
H. schachtii	Beet cyst nematode	Sugarbeet
H. cruciferae	Cabbage cyst nematode	Brassica
H. carotae	Carrot cyst nematode	Carrot
H. punctata	Grass root nematode	Grasses
H. trifolii	Clover cyst nematode	

averaging 0.71 cysts/l of wet sludge in 17 sewage-treatment works in 1982 and 0.03 cysts/l in 54 works in 1984–1985, none of which were viable (McCormack & Spaull 1987). The nematodes are resistant to some sewage treatment processes but numbers are substantially reduced by heated anaerobic processes (DoE/NWC 1981) and cyst viability is affected by mesophilic anaerobic digestion (McCormack & Spaull 1987). Although many empty PCN cysts were found in both cold and heat digested sludges from seven sewage-treatment works over a two-year period, no viable cysts were found (Watson *et al.* 1983, cited by McCormack & Spaull 1987). Cysts of the sugar-beet nematode *Heterodera schachti* in soil samples have been activated by micro-wave irradiation at 650 W for two minutes (Niederwöhrmeier *et al.* 1985).

3.4.3 Effect of sewage and sludge treatments on PCN

A systematic study of the effects of sewage and sludge treatment process has been reported (McCormack & Spaull 1987) on the level and viability of cysts occurring in selected sewage sludges from a particular region (Tayside) of Scotland over a period of one year in 1986. The effects on the viability of both yellow and white PCN of a range of both conventional and novel sewage and sludge treatments were tested in laboratory-scale and/or full-scale operational systems. The methods tested were activated sludge and oxidation ditch sewage treatment, lime treatment of sludge and pasteurization, aerobic thermophilic digestion, mesophilic anaerobic digestion, cold mesophilic anaerobic digestion and accelerated cold anaerobic digestion. Cyst viability was assessed on the basis of motility and configuration of juveniles released from eggs, although in two tests, the number of viable eggs per cyst was used as a criterion.

3.4.4 Incidence of PCN in sludges

Sludges were sampled from five sewage-treatment works which received effluent from vegetable processing, employing oxidation-ditch (2), biofiltration (2) and biofiltration plus activated-sludge treatment processes. Four other works not receiving vegetable processing effluents used oxidation-ditch (2), activated-sludge and biofiltration treatments.

In all, 71 cysts were collected, averaging 0.24 cysts per litre of wet sludge (12 cysts/kg dry solids) of which eight (11%) were viable. Most of the 63 non-viable cysts were empty. Trials with cysts seeded into sludge showed that the efficiency of the cyst recovery techniques used was about 60%, so that the true levels of cyst incidence are probably 50% greater than those reported.

Cyst incidence showed no clear seasonal pattern, although no viable cysts were recovered in spring (February to May). No increase in cyst numbers was found during the period of potato harvesting and washing. Of the 71 cysts, 44 were isolated from the sewage-treatment works receiving vegetable processing effluent, averaging 0.37 cysts/l of wet sludge, and 27 from those that did not (0.18 cysts/l). However, cyst recovery from works receiving vegetable processing effluent was not consistently higher than from those that did not, where comparable treatment processes were used.

3.4.5 Sludge and sewage treatment tests with PCN

During the treatment tests, cysts were held in perforated plastic capsules, which allowed them to come into contact with sewage or sludge.

Over 70% of cysts of *Globodera rostochiensis* remained viable after being stored in tap water at room temperature for 47 days.

3.4.5.1 PCN in mesophilic anaerobic digestion

In operational mesophilic anaerobic digestion at two treatment works, at about 34°C with a mean retention time of three weeks, the number of viable eggs per cyst of *Globodera pallida* showed a significant reduction after three days. Compared with about 300 in the control, the number of viable *G. pallida* eggs per cyst was about 240 after three days and 155 after a week, and was close to none after two and three weeks. However, cysts were found to vary in size, so viability in later results was expressed in terms of the percentage of viable eggs of all the eggs in each test capsule. Tests with *Globodera rostochiensis* showed that the test capsule had no effect on cyst viability, and virtually complete loss of cyst viability of *Globodera rostochiensis* was obtained in three days at one works and after two weeks at the other. The difference between the two works was attributed to differences in sludge composition. Laboratory-scale tests with conditions similar to those in the operational digesters gave similar results, and a significant reduction in the viability of *Globodera rostochiensis* was achieved after three days and virtually complete inactivation after two weeks. *Globodera pallida* cysts were almost completely inactivated after three days.

3.4.5.2 PCN in cold digestion

Capsules of PCN cysts were submerged in sludge held in outside open tanks at two different sites. Viability, expressed in terms of the number of viable eggs per cyst, of *Globodera pallida* was significantly reduced after one week, and decreased to less than 1% after nine weeks. Viability of *Globodera rostocheinsis* was significantly reduced after three weeks, and inactivation was virtually complete after six weeks. The two controls used contained about 300 and 100 eggs per cyst respectively.

In accelerated cold digestion, digestion is accelerated by the addition of mesophilically digested sludge. Viability of *Globodera rostochiensis*, expressed as the percentage of viable eggs, decreased steadily to less than 40% after 15 weeks, then decreased sharply to 2% after 20 weeks, with inactivation virtually complete after 23 weeks.

It was concluded that sludge subjected to the usual period of nine months' digestion should be free of viable PCN cysts.

3.4.5.3 PCN in heat treatments

Pasteurization gave complete inactivation of both *Globodera rostochiensis* and *Globodera pallida* after 30 minutes at 74°C.

Aerobic thermophilic sludge digestion (ATSD), was operated at 60°C with a nominal sludge retention time of five days. With both *Globodera rostochiensis* and *Globodera pallida*, 99.9% inactivation occurred after only one day and complete inactivation after three days.

Clearly, sludge subjected to either of these heat treatments, ATSD and pasteurization, should be free of viable PCN cysts.

3.4.5.4 PCN in lime treatment

Slaked lime was added to raw sludge containing 3.8% dry solids at a rate of 1.28 g/l to raise the pH from 5.5 to 11.5, then PCN cysts were added and left to stand for 24 hours. The treatment reduced the cyst viability, expressed as percentage viable eggs, of both *Globodera rostochiensis* and *Globodera pallida* to about 45%, so that this was not considered an effective treatment for cyst inactivation.

3.4.5.5 PCN in aerobic sewage treatments

Capsules containing cysts of *Globodera rostochiensis* were immersed in the sewage being treated in an oxidation ditch, and although a significant reduction of the percentage of viable eggs was obtained after four weeks, 50% of the eggs still retained viability after seven weeks.

Immersion of *Globodera pallida* cysts in a laboratory-scale activated sludge unit had no effect on the percentage of viable eggs after five days.

It was concluded that neither of the standard aerobic sewage treatment processes is effective in inactivating PCN, and in the cyst incidence survey, viable cysts were recovered from sewage-treatment works operating these processes.

4

Taenia and other parasitic flatworms

4.1 *TAENIA* AND CYSTICERCOSIS

Tapeworms, classed as **cestodes**, are one of the pathogens of major concern in the UK in respect of the agricultural utilization of sewage sludge. The two tapeworms of importance in Europe are the beef tapeworm, *Taenia saginata,* and the pork tapeworm, *Taenia solium,* which occur wherever pork or beef are eaten in an undercooked condition. It is usually the beef tapeworm *Taenia saginata* which is specified in connection with hygiene standards in sewage sludge, although it is infection by the pork tapeworm *T. solium* which is potentially the more dangerous to humans. The most immediate impact is economic, as cattle or pigs infected with *Taenia* are deemed unfit for human consumption, with consequent serious financial loss to the farmer. The loss is exacerbated, because if any infection occurs, it is likely to affect a large proportion of all the cattle grazing on the contaminated land (Bruce *et al.* 1990). Farmers are therefore understandably cautious about the risk of infection by *Taenia* involved with the utilization of sewage sludge, although there is evidence that the transmission cycle between humans and cattle is actually often completed by humans defaecating on to land used for grazing cattle or for growing crops that may be included in animal fodder. A person infected with the tapeworm *Taenia* may excrete a million eggs per day.

Parasitism by *Taenia* also manifests itself by cysticercosis, the development of larval cysts in the muscles, brain or heart. Cysticercosis in cattle is due to *Taenia saginata* and in pigs to *Taenia solium,* but cysticercosis in humans is due only to the larval stage of *Taenia solium,* the pork tapeworm. In the UK, the incidence of cysticercosis amounts to about 0.04% of slaughterings, although light infections may be missed (Bruce *et al.* 1990). The incidence of cysticercosis found in the carcasses of slaughtered cattle in the USA, for example, was 0.026% in 1984, although that in the state of Washington was over seven times higher (official data cited by Hancock *et al.* 1989).

The taeniid worms are classed as **cestodes**, tapeworms, belonging to the phylum of **Platyhelminthes** or flatworms, which also contains the class **trematodes** or flukes.

A tapeworm is segmented and possesses a head, called the **scolex**, with which it attaches to the intestinal mucosa by means of suckers and/or a set of hooks called a **rostellum**. Tapeworms grow by adding segments called **proglottids**, and can attain a considerable length made up of thousands of proglottids, and an indication of tapeworm infection is the presence of these segments in the faeces of the host.

4.2 *TAENIA SAGINATA*

Taenia saginata is the most common taeniid of man, the definitive host, and the adult tapeworm lives only in humans. The larval stage lives in ruminants, especially cattle, as the intermediate hosts. *Taenia saginata* occurs in all countries where beef is eaten.

4.2.1 Life-cycle of *Taenia saginata*

The adult *Taenia saginata* or beef tapeworm has a scolex or head 1 to 2 mm in diameter with four suckers, with which it attaches to the gut wall. The scolex of *Taenia saginata* does not have hooks. The section just behind the scolex is where active cell division occurs and the chain of segments forms. The mature segments are roughly half a centimetre wide and a centimetre or two long. *Taenia saginata* may grow 1000 to 2000 segments and reach a length of typically 5 to 10 m, but sometimes longer, convoluted in the cavity of the bowels. When gravid, each proglottid contains 80 to 100 thousand eggs, and the gravid proglottids separate from the chain to migrate out of the anus or are passed with faeces. In this way, about a million eggs a day are passed by the adult tapeworm. Each segment behaves like an individual worm and has been observed to 'walk' across a petri dish.

Proglottids dry up and burst, releasing embryonated and infective eggs, called **oncospheres**, which are picked up by cattle. The eggs are approximately spherical in shape, with a diameter of 30 to 70 μm. In the duodenum, the egg hatches to release a hexacanth larva, which enters an intestinal venule in order to reach muscle fibres anywhere in the body, where they become parasitic, developing into an infective cysticercus. The larval worm is called *Cysticercus bovis*. A person who eats infected beef, which is cooked insufficiently to inactivate the cysticerci, also becomes infected, since the cysticercus evaginates into an adult tapeworm. The tapeworm reaches maturity with about ten weeks, and can live in the human intestine for as long as 25 years (Feachem *et al.* 1983, Ketchum 1988, Jeffrey & Leach 1975).

4.2.2 Epidemiology of *Taenia saginata*

The life-cycle of *Taenia saginata* requires that humans must ingest a viable cysticercus by eating contaminated beef that has been insufficiently cooked. The adult tapeworm lives only in humans. The life-cycle is completed by cattle eating food contaminated with faeces from an infected human. Human infection is thus highest in parts of the world where beef is a major food and sanitation is poor. It tends to occur in communities where humans and cattle live in close proximity, and is exacerbated by poor sanitation. There are some areas, however, where the incidence of cysticercosis in cattle is as high as 10%, while the adult tapeworm occurs in only one human in ten thousand (Feachem *et al.* 1983).

There has been some difficulty in explaining why infection with *Taenia saginata* is still fairly common in developed countries with proper sanitary facilities, and why its occurrence has increased since 1945. A possible explanation is that, even in developed countries, cattle are raised in country areas where sanitation may be unsophisticated or absent. Even where sanitary facilities are available, a cowhand out in the fields with an urgent need to defaecate is more likely to defaecate on the pasture than to travel several kilometres to the nearest sanitary facility. In addition, it has become fashionable to eat beef undercooked, rare or 'blue'. Apart from this, suspicion has fallen mainly on the use of sewage sludge in agriculture, and also transmission from sewage treatment works to pastures by birds and insects. *Taenia* parasites are a problem in many developing countries, especially where night-soil is spread on land. Infection can be prevented by cooking meat well or by freezing at −5°C for at least a week.

4.2.3 Pathogenesis of *Taenia saginata*

The presence of the *Taenia saginata* tapeworm in humans is generally almost symptomless, but verminous intoxication caused by absorption of the excretory products of worms may occur with dizziness, abdominal pain, headache, localized sensitivity to touch, nausea, hunger pains and loss of appetite. Although the larval stage produces an immune response in cattle, humans have not been found to develop immunity against the adult tapeworm. The development of immunity by the intermediate host means that the infective dose, the number of eggs that the animal has to ingest for infection to occur, depends on whether or not the animal has previously been infected. As humans have no apparent immune response to the cysticerci, then, in principle, only one larva is needed to form a tapeworm.

4.3 *TAENIA SOLIUM*

The pork tapeworm, *Taenia solium,* is the most dangerous of the adult tapeworms. Although *Taenia solium* has many similarities to *Taenia saginata,* there are important differences in its infection risk to humans. Like *Taenia saginata,* the adult stage of *T. solium* lives only in humans, but while the larval stage usually lives in pigs, it can also infect humans and other domestic animals.

4.3.1 Life-cycle of *Taenia solium*

The scolex of the pork tapeworm has two circles of 22 to 32 non-retractable hooks, called a **rostellum** as well as four suckers, unlike *Taenia saginata,* which has only suckers. The life-cycle of *Taenia solium* is very similar to that of *Taenia saginata,* except that the intermediate hosts are usually pigs, and not cattle, and the strobila, the chain of proglottid segments, is typically 3 m long and rarely more than 10 m long. Gravid proglottids passed in faeces are laden with eggs which are infective to swine. When eaten, the oncospheres develop into cysticerci in the muscles and other organs. The cysticercus evaginates and attaches to the small intestine and matures in five to 12 weeks. Pathogenesis in the animal host is similar to that of *Taenia saginata,* with the development of cysticercosis in the muscles of the pig after nine or ten weeks. The larval form in the pig is called *Cysticercus cellulosae.*

4.3.2 Epidemiology of *Taenia solium*

Humans become infected by eating pork infected with *C. cellulosae,* or, unlike infection by *Taenia saginata,* by ingesting eggs of *Taenia solium* in food, water, or from people who are infected with the adult tapeworm. Infection through the mouth by a carrier of the tapeworm is called *autoinfection.* Humans can thus act as an intermediate host, as well as the definitive host. It is also considered possible that autoinfection can occur by eggs being carried from the intestine back to the stomach, where they hatch, although there is some doubt as to whether this actually occurs (Feachem *et al.* 1983).

4.3.3 Pathogenesis of *Taenia solium*

Unlike most other species of *Taenia,* the cysticerci of *Taenia solium* will develop readily in humans and cause cysticercosis. Infection occurs when eggs hatch in the stomach. *Cysticercus cellulosae* is about a centimetre in size, and cysticerci may occur anywhere in the body, usually in several places simultaneously, but especially in the muscles and the brain, with very serious consequences. A common symptom is epilepsy and occurs where cysticerci are found in the brain. When a cysticercus dies it elicits a severe inflammatory response which may prove fatal to the host. Prevention of cysticercosis depends on elimination of adult tapeworm and a high level of personal hygiene.

4.4 TAENIID OVA

The eggs of *Taenia saginata* and *Taenia solium* are identical, and separation of the species cannot be made by studying ovum morphology. The ovum of *Taenia* spp. is dark brown, round or slightly ovoid, and approximately 35 μm in diameter. All taeniids belong to the phylum Platyhelminthes, class Cestodea, and have a common egg structure. At the centre of a fully developed *Taenia* egg is the oncosphere or hexacanth embryo, which is infective to the intermediate host and has three pairs of delicate hooklets each about 15 μm long. The oncosphere is immediately surrounded by an embryophore which consists of prism-shaped blocks that give a visual impression of radial striations. The capsule is the outermost covering which is lost when *Taenia* ova appear in faeces or sewage.

The egg production capacity of a single *Taenia* tapeworm is vast, since a single gravid proglottid may contain up to 95 000 eggs, while an adult worm could produce 600 million eggs per year.

4.4.1 Viability and hatching of taeniid ova

A survey of available literature indicates a need for a universally accepted definition of a viable taeniid ovum. Silverman (1954) reported that hatching of the oncosphere resulted from disintegration of the embryophore, while activation of the six-hooked embryo to tear its way out of the oncospheral membrane followed only after hatching occurred. Silverman concluded that 'the general appearance of the oncosphere is an indication of its viability. The cytoplasm of dead or dying oncospheres is visibly more granular than that of living specimens. Non-viable oncospheres also lose their characteristic egg shape and become ovoid' and found that a dead or dying oncosphere took up Nile Blue stain more readily than a living one. Work at the Liverpool School of Tropical Medicine (Owen 1983) suggested that staining of viable ova with tetrazolium salts after treatment with bile and pepsin provided a positive and conclusive test of *Taenia* ova viability.

Webbe (1967) provided a detailed review of the hatching and activation of taeniid ova in relation to the development of cysticercosis and man, and concluded that the ova of both *Taenia solium* and *Taenia saginata* required a protease for the disintegration of the embryophore and hatching of the hexacanth embryo prior to its activation by tryptic digestion together with the influence of bile salts in the intestine. A new method for the *in vitro* hatching and inactivation of *Taenia taeniaformis* was provided by Brandt and Sewell (1981), which involved pre-treatment with hypochlorite at 0.67% w/w available chlorine, to cause disaggregation of the embryophore blocks, followed by exposure to a solution of trypsin, ox-bile and calf serum in HEPES buffer. Gallie and Sewell (1970) found the *Taenia* hatching technique of Silverman (1954) of little value, since only about 5% of the embryos hatched. They found that carbon dioxide was an important stimulus in hatching, and devised a modified technique whereby the addition of concentrated hydrochloric acid produced a reaction with sodium bicarbonate, resulting in the production of carbon dioxide and a consequent 40 to 47% hatch.

4.5 *ECHINOCOCCUS GRANULOSUS*

Echinococcus granulosus, also belonging to the phylum Platyhelminthes (flatworms), the class Cestodea (tapeworms) and the family Taeniidae, is a very small taeniid tapeworm, with the adult measuring 3 mm to 6 mm long. However, the larval forms are large and can infect humans, resulting in hydatidosis, hydatid disease. Carnivores, in particular dogs, are the intermediate hosts for *Echinococcus granulosus* and herbivorous animals are the definitive hosts. The intermediate hosts become infected by eating eggs on contaminated herbage, and humans may become intermediate hosts by accident.

Hatching and migration of the *Echinococcus granulosus* oncosphere is similar to that of *Taenia saginata,* except that the liver and lungs are the commonest site of development. The oncosphere grows into a unilocular hydatid cyst which contains brood capsules. The brood capsules are small cysts, each containing ten to thirty protoscolices which later mature into adults.

Dogs are infected with *Echinococcus granulosus* when they eat sheep offal or butchered animals, while herbivores are infected after ingesting herbage contaminated with dog faeces. The possibility that herbage may be contaminated with sewage containing *Echinococcus granulosus* ova should also be considered. Humans tend to become infected after acquiring ova by fondling pet dogs.

Symptoms of *Echinococcus granulosus* infection often takes twenty or more years to appear and are related to increase in size of the hydatid cyst which affects neighbouring organs. Hydatid cysts tend to develop in the liver or brain, and a ruptured hydatid cyst will lead to anaphylactic shock and death. Occasionally, cysts can develop in the bones, with the consequent bone damage leading to spontaneous fractures.

Treatment of hydatid cysts involves surgery where the cyst capsule is incised and the contents aspired followed by injection of 10% formalin to kill the cyst germinal layer.

A similar organism, *E. multilocularis,* has a normal life-cycle involving a wild carnivore, such as a fox, as the host and rodents as the intermediate hosts. Humans may become accidental intermediate hosts by ingesting ova, for example on fruit contaminated with fox droppings (Jeffrey & Leach 1975) The life-cycle is then similar to that of *E. granulosus.* The distribution of the organism is given as principally north America and eastern Europe.

4.6 *TAENIA MULTICEPS*

Carnivores, especially dogs and foxes, are the definitive hosts of *Taenia multiceps,* and ungulates, such as sheep, the intermediate hosts of *Taenia multiceps,* which belongs to the order Cyclophyllidea and the family Taeniidae. The larval stage of this parasite is the coenurus type or bladder-worm, which is similar to a cysticercus, but has many scolices. Coenuriasis of sheep causes a characteristic vertigo called **gid** or **staggers**. Coenurus occasionally develops in man, affecting the brain, eye, muscles and subcutaneous tissue. It is likely that *Taenia multiceps* ova may occur in sewage sludges, but is not expected to feature frequently in effluent parasitology.

4.7 *DIPHYLLOBOTHRIUM LATUM* OR *DIBOTHRIOCEPHALUS LATUS*

Diphyllobothrium latum and *Dibothriocephalus latus* are synonyms for the fish tapeworm, which has humans, dogs and cats as its definitive hosts and two aquatic intermediate hosts. Eggs are passed in the faeces of their mammalian hosts, and if passed into water, hatch into free-swimming larvae called coracidia which are ingested by miniscule aquatic creatures called cyclops. These in turn are consumed by freshwater fish, and the larvae become established in the muscles of the fish. The life-cycle is then completed when humans or the other hosts eat the infected fish raw or undercooked. The tapeworm can grow several thousand proglottids to give total lengths of 3 to 10 m, but there are no pathological effects, except occasional vitamin B12 deficiency, due to absorption by the worm.

4.8 *MONIEZA SPECIES*

Two similar tapeworms, *Monieza expanza* and *Monieza benedeni,* are parasites of sheep, goats and cattle, with a mite as an intermediate host in the *Monieza* life-cycle. The eggs are triangular to square in shape, about 50 to 75 μm in diameter, and contain an oncosphere with the hexacanth embryo, surrounded by an embryophore. A gravid segment of the tapeworm is passed in the faeces, from which eggs are liberated and eaten by a mite of the genus *Oribatidae.* A cysticercoid larva is produced within the mite, which infects ruminants when they eat material containing mites and larvae, thus completing the cycle. Heavy infection by *Monieza* spp. may lead to weight loss in ruminants, but otherwise these tapeworms are not highly pathogenic. It is not unreasonable to expect *Monieza* ova to be disseminated in inadequately treated sewage sludges applied to farmland.

4.9 CYSTICERCOSIS AND SEWAGE

Taenia saginata is probably the only human helminth parasite regularly transmitted through sewage in the UK, although many of the protozoan and helminth parasites now occurring outside the tropics could in theory do so (Owen & Crewe 1985). The application of sewage sludge to grazing land provides an economic and efficient alternative to coastal discharge as a means of disposing of sewage sludge. It is likely that this practice also provides a potential route for the transmission of *Taenia saginata* eggs from human faeces to cattle, in addition to the routes provided by birds, invertebrate animals, streams and surface water.

Bovine cysticercosis may be defined as the infection of cattle with larval stages of the human beef tapeworm *Taenia saginata*. The UK Public Health Laboratory Service reported 82 cases of human tapeworm infection in 1976, 98 in 1977 and 58 in the first 32 weeks of 1978 (Crewe & Owen 1978). A review of published data for the USA by Snowdon *et al.* (1989a) cited a 1980 estimate that the incidence of *Taenia saginata* infection in the USA was less than 1%, and a 1974 survey of US public health laboratories found evidence of adult taeniid tapeworm infection in 0.05% of over 400 000 human faecal specimens examined.

Many authors have drawn attention to the presence of *Taenia* eggs in sewage and to the possibility of ova dissemination when sewage is applied to agricultural land (Silverman & Griffiths 1955, Gemmell & Johnstone 1977, Crewe & Owen 1978, Wilkens 1982, Bürger 1983, Lewis-Jones & Kiff 1984, Snowden *et al.* 1989a, Bruce *et al.* 1990, for example), and application of sewage sludge to pasture has been held responsible for an outbreak of bovine cysticercosis (MacPherson *et al.* 1978, 1979). In sludges treated by cold or mesophilic anaerobic digestion from seven sewage-treatment works, *Taenia* ova were found at a level averaging 11 in a range of 0 to 16 ova/l (Watson *et al.* 1983).

Three pathways have been outlined for human infection with *Taenia saginata* (Reilly *et al.* 1981):

(1) through direct or close contact between human faeces and cattle,
(2) through an intermediate vector, such as birds or insects,
(3) through bovine contact with human sewage, in the form of either wastewater or sludge.

Arundel and Adolph (1980) observed removal of *Taenia* ova by various treatment processes, and, from a survey of cysticercosis in cattle grazed on grass irrigated with effluent (Table 4.1), showed that raw sludge did transfer viable ova to land, while none of the cattle grazed on pasture irrigated with lagooned sludge was infected.

Jepsen and Roth (1950) studied the distribution of *Cysticercus bovis* in calves fed on pasture irrigated with sewage for 4½ to five months. After slaughter, all 20 calves investigated were found to be infected with *C. bovis* with the number of cysts observed per individual ranging from two to 31. The authors concluded that using sewage for irrigating and fertilizing fields involved public health hazards.

Table 4.1 — Number of *Cysticercus bovis* found in calves feeding on land treated with wastewaters (Lewis-Jones 1984 from Arundel & Adolph 1980)

Treatment	No of calves inspected	Incidence of infection
Raw sewage	40	30.0%
Activated sludge	33	9.0%
Trickling 'filter'	30	3.3%
Lagooning	40	0
Controls	40	2.5%

Crewe and Owen (1979) observed a definite pattern in cysticercosis caused by sludge disposal. They believed that the spreading of sewage sludge provided an even, low-level contamination of pasture with *Taenia saginata* eggs and that heavy contamination required a contamination of eggs caused by segments dispersed with sludge or bird droppings.

Between 1977 and 1979, there were five outbreaks of cysticercosis in cattle associated with the application of sewage sludge to agricultural land. Reilly and Collier came to an interim conclusion 'that there is a relatively low correlation between the use of sewage sludge and cysticercosis if the guidelines . . . are followed'. They further confirmed their belief that incidence of bovine cysticercosis does not correlate directly with sewage sludge application. Of all the 21 381 farms studied, bovine cysticercosis was found on 370, an incidence of 1.7%. Of 339 farms taking sludge, 16 had bovine cysticercosis, an incidence of 4.7%. The risk of bovine cysticercosis was therefore three times greater with sludge usage than without, and it was noted that two-thirds of the sludge used was digested. In reviewing the final report by Reilly *et al.*, published in 1986, on the results of their regional survey of bovine cysticercosis in 1980 to 1983, Bruce *et al.* (1990) concluded there was a significant association between cysticercosis and the use of sludge. However, less than 5% of the affected farms had used sludge, so that routes of infection other than with sludge are important (Bruce *et al.* 1990). Nevertheless, such occurrences emphasize the importance of carefully controlled sludge treatment and disposal practice to ensure that the eggs are non-infective for cattle and to maintain assurance that sludge usage is safe. In addition, *Taenia saginata* eggs are considered to lose viability and infectivity completely after six months in sludge or soil (Pike & Davis 1984). Laboratory experiments showed that viability declined at 0.24% per day in saline, which also indicates complete loss of viability after about six months, but eggs were rendered non-viable in three hours at 55°C in sludge.

Thus while the contamination of land with eggs of *Taenia saginata* can arise only from faecal pollution, the use of sludge is only one possible means of faecal contamination. A recent survey of the pattern of cysticercosis, in data from three slaughterhouses and 18 feedlots in Washington, USA, indicated human faecal contamination of cattle feed (Hancock *et al.* 1989). The most likely source of *Taenia saginata* ova was considered to be potato processing waste, as the timing of infection peaks corresponded to the use of waste from freshly produced, rather than stored, potatoes, after allowing for the delays resulting from transport and organism development. This presumably occurred as a result of human defaecation on to fields where potatoes were grown.

In a review of bovine cysticercosis encountered on meat inspections in England and Wales between 1969 and 1978, it was seen that the figures for the last two quarters of the year were higher than those from the first two quarters (Table 4.2). This correlated with the time of year when animals were grazing and therefore exposed to *Taenia saginata* ova (Blamire *et al.* 1980).

Table 4.2 — Seasonal incidence (as percentage of total kill) of bovine cysticercosis (Lewis-Jones 1984 from Blamire *et al.* 1980)

Year	Quarter			
	First	Second	Third	Fourth
1969	0.165	0.102	0.251	0.208
1972	0.062	0.073	0.108	0.106
1976	0.036	0.042	0.053	0.052
1978	0.028	0.034	0.064	0.046

The prevalence of cysticercosis in cattle appeared to be decreasing until the early 1980s (Crewe 1983), but has remained constant at about 0.04% of slaughterings for the last ten years (Bruce *et al.* 1990). The observed pattern of cysticercosis is possibly due to a combination of several factors, including pasture contamination level and genetic, immunological and biochemical differences in cattle. The economic loss to the meat trade caused by bovine cysticercosis is not confined to the UK, as it causes a considerable loss of meat in Australia in both the home and export markets (Arundel 1972). Once on pasture, *Taenia saginata* ova viability and survival are dependent on environmental conditions. Silverman (1954) concluded that the maturation of taeniid ova outside the proglottid was temperature-dependent, while Gemmell and Johnstone (1977) drew attention to the fact that taeniid eggs may survive for long periods at temperatures below 38°C and that ultra-violet irradiation appeared to reduce ova infectivity.

4.10 SLUDGE TREATMENT AND *TAENIA SAGINATA* OVA

Investigations of the resistance of helminth eggs to wastewater and sludge treatment processes tend to use the eggs of *Ascaris* spp. as test organisms. This is partly because they are easier to obtain than eggs of *Taenia* spp., and also because they are generally found to be more resistant to treatment conditions. *Ascaris* spp. provide excellent biological models of nematodes, and the high resistance of *Ascaris* eggs makes them good test organisms, as a treatment designed to inactivate eggs of *Ascaris* spp. should then deal with eggs of *Taenia* spp. equally well or better. However, Owen and Crewe

(1985) warned that it is difficult, and often confusing, to compare the effects of chemicals on *Taenia* eggs with those on *Ascaris* eggs, as the eggs differ considerably in morphology and structure, and it is an over-simplification to consider that *Ascaris* eggs are more resistant to chemical treatment and other adverse conditions than *Taenia* eggs. Chemicals effective against *Taenia* can be less so against *Ascaris* and vice versa. There is thus a need to investigate the survival of *Taenia*; extensive work on *Taenia* ova has been carried out by Crewe *et al.* at Liverpool SHTM and they are the main source of *Taenia* eggs in the UK. The effects of temperature on *Taenia* eggs is summarised in Fig. 4.1.

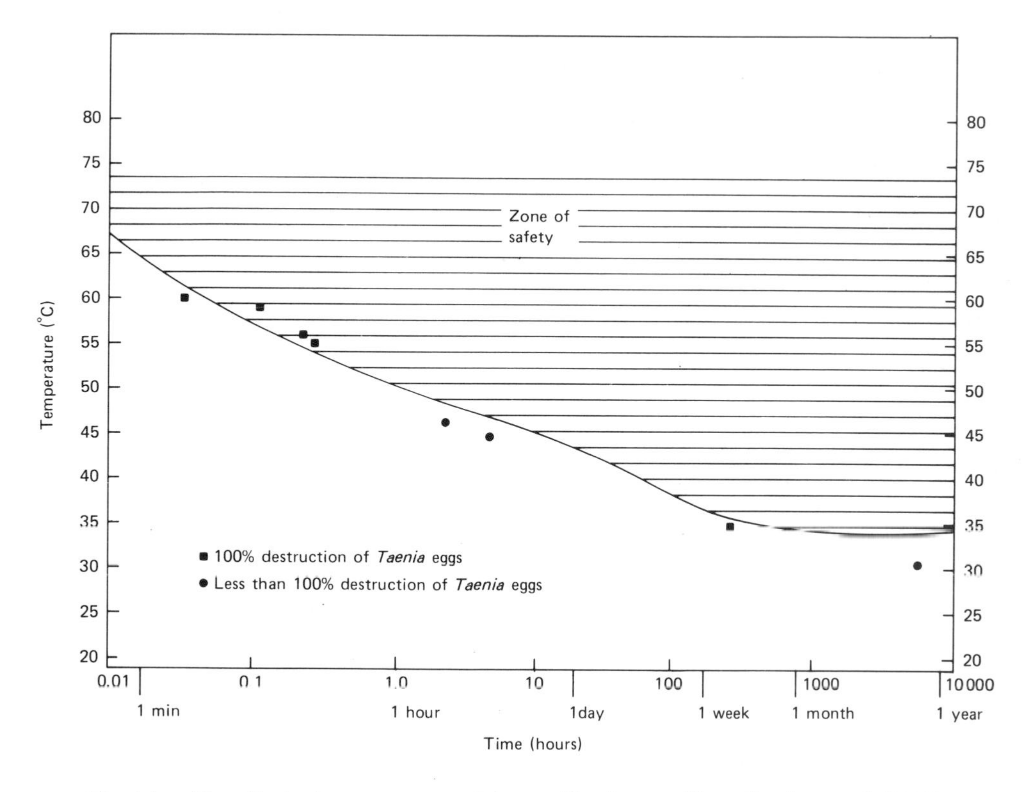

Fig. 4.1 — The effects of temperature and time on *Taenia* eggs. (From Feachem *et al.* (1983). Reproduced by kind permission of the World Bank.)

4.10.1 Effect of proglottids

Studies investigating the effects of various treatments, such as digestion, lagooning, de-watering and/or composting, on the viability of tapeworm eggs tend to use eggs freshly dissected from proglottids. Although eggs are shed from proglottids within the human alimentary canal, eggs are found in faeces still within proglottids, and, following chemical purging of the gut, even within whole tapeworms, which then enter the sewage treatment process. The egg may then be protected by the proglottid from the effects of the sewage treatment process and when exposed on pasture.

A study by Storey (1987) demonstrated that eggs within proglottids survived longer than freshly dissected eggs during sludge storage, and laboratory-scale aerobic and anaerobic mesophilic and thermophilic digestion of sludge. For killing *Taenia saginata* eggs, anaerobic digestion was found to be more effective than aerobic digestion and lagooning. In all processes, temperature was the most important factor, with the rate of egg-kill increasing with increasing temperature, and at 55°C, eggs both free and within proglottids were killed within a few hours. Comparison with other reported results is difficult because of the change in resistance of eggs to adverse conditions as they mature, within or free of the proglottid. Earlier results indicated destruction of *Taenia saginata* eggs after 56 days at 26 to 28°C, or 5 days at 35°C, or that *Taenia saginata* eggs survived up to 26 days' digestion at 30°C with death rates of 0.24%/day at 39°C in saline, and very rapid deaths at temperatures greater than 50°C.

Förstner (1968) subjected ascarid and *Taenia* eggs to anaerobic digestion in sludge at 30°C for 26 days. The sludge containing ova was then applied to land. After five months' outdoor exposure, most of the ascarid eggs survived, while only a small percentage of *Taenia* eggs isolated from proglottids remained viable.

4.10.2 Sludge treatments and infectivity of ova

There is clearly difficulty in specifying a useful and reliable criterion for determining when a helminth egg has been inactivated. The most important criterion is really whether cattle become infected by ingesting parasite eggs which have been subjected to a defined treatment process. Such an investigation has been carried out by the AFRC Animal Disease Research Centre and the former Severn-Trent Water Authority (Hughes *et al.* 1985) and at the Water Research Centre, Medmenham (Bruce *et al.* 1990).

The effects of various sludge treatments on reducing the infectivity of *Taenia saginata* eggs were tested by Hughes *et al.* (1985). The viability of the *Taenia saginata* eggs was tested by infecting calves with them and comparing the cysts produced after three months in calves fed with treated eggs with those in calves fed with untreated eggs in equal numbers. The occurrence and distribution of cysticerci were determined in post-mortem examinations. Viability tests carried out *in vitro* did not correlate with infectivity.

Some treatments were carried out as batch operations, with a defined treatment time, while others were carried out as intermittent or continuous operations, with a range of treatment times.

Compared with untreated eggs of the same age, *Taenia saginata* eggs in sludge treated at 55°C for three hours had their infectivity decreased by over 99%, and infectivity was completely destroyed by anaerobic digestion for ten days at 35°C followed by lagooning for 15 days. Lime treatment to give pH 12 reduced egg viability by 96%, aerobic digestion for 16 days by 82% and lagooning for 28 days by 99%. Hughes *et al.* concluded that mesophilic anaerobic digestion is very effective in decreasing the infectivity of *Taenia saginata* eggs, other treatments commonly used in smaller treatment works give a significant reduction in infectivity and that pasteurization conditions of 30 minutes at 70°C provide a considerable safety margin.

After allowing for the differences between batch, intermittent and continuous operation, the results were summarized graphically by Bruce *et al.* (1990) as a ranking table, according to the ratio of cysts produced by treated eggs to those by untreated eggs (treated/untreated), and are expressed here numerically (Table 4.3).

Table 4.3 — Ratio of cysts produced for treated/untreated sludges (data from Bruce *et al.* 1990)

Treatment	Cyst ratio
Lagooning for 28 days at 7°C	0.002
Pasteurization, 3 hours, 55–60°C plus anaerobic digestion 30 days, 35°C	0.003
Anaerobic digestion, 30 days, 35°C plus lagooning 15 days, 7°C	0.005
Pasteurization, 3 hours at 55°C	0.013
Pasteurization, 3 hours 60°C	0.013
Lime treatment, pH 12 for 24 hours	0.03
Pasteurization 3 hours, 52°C plus anaerobic digestion 24 hours, 35°C	0.04
Lime treatment, pH 12 for 48 hours	0.05
Thermophilic aerobic digestion, 6 days at 50°C	0.055
Accelerated cold digestion for 50 days at 3°C	0.074
Anaerobic digestion for 15 days at 35°C	0.11
Anaerobic digestion for 30 days at 35°C	0.15
Aerobic digestion for 16 days at 7°C	0.27
Accelerated cold digestion for 50 days at 3°C	0.34

The effects of other treatment processes on *Taenia saginata* reported in the same volume as those of Hughes *et al.* (1985) were summarized by Havelaar (1985). The infectivity of eggs of *Taenia saginata* was greatly decreased by conventional mesophilic digestion of about one month at about 30°C, and by lagooning sludge for one month. *Taenia saginata* eggs are inactivated by lime treatment giving pH 12.0 for 24 hours. With other chemical treatments, inactivation of *Taenia saginata* eggs can be due to inhibition of hatching by lime, alkaline formalin and quaternary ammonium compounds; or by direct ovicidal action by bis-biguanides, copper ions or a combination of both, or by acid formalin or peracetic acid.

Data presented in graphical form by Bruce *et al.* (1990) showing the treatment times required to achieve a 90% reduction in viability for infectivity of *Ascaris suum, Taenia saginata* and other pathogens and faecal organisms in sludge at different temperatures in the range 40 to 80°C indicated that the examples of effective sludge treatments given in the 1989 Code of Practice are generally more than sufficient to give 90% destruction, except that the viability of cysts of the potato cyst nematode is not destroyed by lime treatment. Cold digestion, aerobic stabilization at ambient temperature, did not give 90% reduction in *Salmonella* numbers or *Taenia saginata* infectivity, but would achieve this if followed by storage for four weeks.

An earlier evaluation of the effect of anaerobic sludge digestion on *Taenia saginata* ova, at temperatures between 4°C and 60°C, showed that temperatures below the 35°C normally used for mesophilic digestion would have little effect in reducing viability of *Taenia saginata* ova (Pike *et al.* 1983), while temperatures above 40°C gave an effective reduction in viability. Experiments involving micro-straining with stainless steel gauze and sand filtration showed that the latter was able to remove 90 to 95% of *Taenia* eggs, while the former method was far less effective (Silverman & Griffiths 1955).

4.11 CHEMICAL INACTIVATION OF *TAENIA*

Some conventional sludge treatments do not destroy helminth ova or sludge odour, and restrictions on grazing after land application are unpopular with farmers, and therefore likely to be flouted. This has aroused interest in the use of chemical disinfectants to limit the transmission of parasites, such as *Taenia saginata*, and contamination of pastures with pathogens such as *Salmonella*. Doubt has been expressed that application of sludge to pasture is in fact a cause of *Cysticercus bovis* infections. *Ascaris suum* is able to infect lambs grazing on land on which pig slurry has been disposed, rendering their otherwise valuable livers unsaleable (Owen 1984).

Chemical treatment appears to provide an economically attractive means of disinfection of sewage sludges, and several chemicals have shown promising ovicidal activity (Owen & Crewe 1985). Only a minority of commercially available disinfectants appear to have a marked effect on *Taenia* eggs, and very few could be used with cost-effective concentrations or contact times. In addition, the effects of these compounds on the land and animals thereon when the chemically disinfected sludge is spread must be questioned. Lime and formalin-based mixtures inhibit hatching of *Taenia* eggs, but appear to leave the embryo unharmed. Quaternary ammonium compounds, peracetic acid and bis-biguanides affect the viability of the embryo, and some compounds containing copper show considerable ovicidal action.

The effects of selected chemical disinfectants were reported by Owen (1984). To determine the viability of eggs of *Taenia saginata* it is necessary to remove the outer embryophore and stimulate the embryo to emerge from the oncospheral membrane. With *Ascaris suum* eggs, the procedure is far simpler because they undergo a series of easily identifiable development stages between leaving the female worm and infecting the next host. Chemical disinfectants were assessed by comparing treated and untreated batches of *Taenia saginata* and/or *Ascaris suum* ova. The ovicides tested were lime added to water, sludge-liquor and sludge to give pH 11.5; acidified formalin (SX-1); alkaline formalin (DF 955 and a modified formulation DF Mod); peracetic acid (PAA); bis-guanide (VAN-1, VAN-2 and VAN-3) and quaternary ammonium (V-CL and 739) compounds. Trials have also shown that *Taenia hydatigena* eggs respond to lime and formaldehyde-based and bis-guanide disinfectants identically to those of *Taenia saginata,* but *Taenia pisiformis* eggs do not. Tests of the effects of lime were reported by Poole and Mills (1984), and tests of the effects of lime, DF 955 and peracetic acid on *Taenia saginata* were also reported by Godfree *et al.* (1984): these are summarized further on.

Owen (1984) reported that the effects of lime on *Taenia saginata* eggs were that hatching was completely inhibited and no viable embryos were produced following exposure to lime in water (four hours at pH 11.5) and DF 955 (24 hours with 1.0% v/v). Hatching was completely inhibited by lime in sludge-liquor (24 hours at pH 11 or three hours at pH 12). Lime in sludge (24 hours at pH 11.5) gave only 10% inhibition of hatching but only 44% of the eggs produced viable embryos.

Results with peracetic acid (PAA) were found to be anomalous by Owen (1984), as 100 mg/l PAA completely inhibited hatching after exposure for 30 minutes and 24 hours, but after seven days exposure, all eggs hatched from which 78% of the embryos were viable. Thirty minutes' exposure to 500 mg/l PAA gave complete inhibition of hatching, whereas all eggs hatched after 24 hours' exposure and 85% of the resulting embryos were viable. The bis-guanide and quaternary ammonium compounds had little effect on the proportion of eggs hatching, but the proportions of viable embryos resulting was generally about 30% for 4% ovicide solutions in water and about 20% for 4% ovicide solutions in sludge-liquor and 20% ovicide solutions in water and sludge-liquor. An ovicide designated MB 724 produced complete inhibition of hatching and lysis of *Taenia saginata* eggs at concentrations as low as 62.5 ppm in water and 300 ppm in sludge-liquor.

An attempt was made by Owen (1984) to correlate *in vitro* viability with *in vivo* infectivity, by comparing animals given untreated eggs with those given eggs treated with lime (pH 12.03 for three days), DF 955 (10% for 18 hours) and VAN-1 (20% for one hour). Treatment with lime and with DF 955 gave virtually complete inhibition of hatching, while large numbers of mature cysts were recovered from the animals dosed with the VAN-1-treated eggs. Comparison with the control group was complicated by the possibility that doses of nearly 1000 high-viability eggs triggered an overwhelming immune response.

The effects of chemical disinfectants on *Ascaris suum* eggs were that lime treatment (pH 11.5 for 48 hours) left only 1% reaching the larval stage in sludge, but 63% in water. Treatment with DF 955 for 39 days in water gave 1% and 0% ova in larval stage with 4.0% and 10% DF 955 respectively. No *Ascaris suum* ova were in the larval stage after exposure to 20.0% 739 in water for 35 days.

The effectiveness of chemical disinfectants on *Salmonella* and *Taenia saginata,* the organisms of greatest concern that might be transmitted by sewage sludge, was assessed in a study by Godfree *et al.* (1984).

Disinfectant tests reported by Godfree *et al.* (1984) were on DF 955, on an alkaline formulation containing formaldehyde and oil, and Proxitane 4002, containing 36 to 40% w/w peracetic acid (CH_3COOOH) with some hydrogen peroxide and acetic acid. Their action on several serotypes of *Salmonella* inoculated into raw and anaerobically digested sludges was investigated in both laboratory and field trials.

At a concentration of 1% v/v in sludge and with a contact time greater than one hour, DF 955 destroys very high levels of *Salmonella,* gives 68% inactivation of *Taenia saginata* ova and eliminates sewage sludge odour, with no evidence of adverse effects on soil invertebrates or plant species. A concentration of 1 g/l peracetic acid in sludge, using Proxitane 4002, is significantly bactericidal and ovicidal, giving over 98% reduction of *Salmonella* in both raw and digested sludge in field trials with a one-hour contact time, and in laboratory trials, 99.9% reduction of *Salmonella* levels in raw sludge with a contact time of only ten minutes. Major reductions in the viability of *Taenia saginata* ova were also observed using these levels of peracetic acid.

Lime is the most convenient and readily available chemical for use as a disinfectant, as it is widely used as a sludge conditioning and de-watering aid. Lime treatment is also attractive because it does not require expensive and/or sophisticated plant, and so is suitable for small treatment plants. If, in addition, exposure times of only a few hours after lime-treatment were found to inhibit parasite hatching, then the road tanker itself could be used as the lime-treatment vessel, with the following transit time sufficient exposure to achieve disinfection. The effect of high pH levels on *Salmonella* viability reported by Godfree *et al.* (1984) was that, with a contact time of 15 minutes, pH values above 9 produced a marked reduction in the *Salmonella* count, and a 99% reduction was achieved at pH 11.5, at an unspecified temperature. Exposure of eggs of *Taenia saginata* to lime water over a range of pH values indicated that four hours at pH 12 were needed to kill 90%. After four hours at pH 11.5, hatching of *Taenia saginata* ova was reduced from 35% to 0.6%, and to 0.75% after two hours at pH 12, at an unspecified temperature.

The effect of lime treatment was tested by Poole and Mills (1984) on ova of *Taenia saginata* and on a related species, *Taenia hydatigena,* of which the intermediate *Cysticercus tenuicollis* is found in the liver and connective tissue of sheep and the adult tapeworm in the intestines of dogs.

Taenia ova were suspended in saline as a control and in three types of liquor, from raw mixed sludge (pH about 6), and from sludges treated with lime to give pH 11 and pH 12 respectively, for periods of time from a few hours up to 11 days. With *Taenia saginata* ova, complete inhibition of hatching occurred after three hours at pH 12 and 24 hours at pH 11, and a considerable reduction in successful hatching occurred in raw liquor after three hours. There was no evidence that lime treatment at either pH 11 or pH 12 affected the viability of those ova that succeeded in hatching. The effect of liming was considered to be due entirely to the effect of pH, presumably due to free ammonia formation, and not to calcium incorporation.

With *Taenia hydatigena* ova, at pH 12, hatching was strongly inhibited after 1 hour 40 minutes and completely inhibited after 44 hours. After eight days' exposure, hatching was strongly inhibited in raw liquor, but in liquor at pH 11, 33% of ova hatched. While this implies that raw sludge liquor was more inhibitory to *Taenia hydatigena* than to *Taenia saginata,* this could be due to a difference in the freshness of the original ova used. In lamb trials with *Taenia hydatigena,* lime-treated ova produced only one-fifth of the level of infection compared with the control (saline) ova.

The above results indicate that lime treatment has little effect on the viability of *Taenia* ova, but does inhibit their hatching and so reduces the risk of infecting livestock.

In a discussion of these reports (Bruce 1984), Mr K. Guiver said that he believed the main factor in killing *Taenia saginata* ova in sludge digestion was in fact the anaerobic conditions, with a redox potential below −275 mV, and asked why they were not killed in the anaerobic conditions of the rumen. One of the present authors queried the difference between 'viability' and 'infectivity', because infectivity can be assessed only in animals, by cyst production. Mr D. C. Watson and Mr Poole said they thought that the ammonia liberated at high pH levels was the active factor in killing pathogens.

4.12 *FASCIOLA HEPATICA*

Fasciola hepatica, the 'liver fluke' or 'sheep liver fluke', is a trematode and occurs mainly in goats, sheep and cattle, causing 'liver rot'. Other mammalian hosts may become infected and, man for example, may become infected through eating raw liver from infected animals or watercress which contains metacercariae. Human infection is, however, uncommon, so that sewage and sewage sludge is unlikely to contain the eggs of the fluke, unless it has been mixed with manure from infected animals.

Fasciola hepatica is a large fluke up to 30 mm long and 13 mm wide. It has a characteristic leaf-like shape with an oral sucker. There is an anterior acetabulum and the tegument is covered with spines. The morphology and structure of the liver fluke is well known and it is used as a biological model for the *Trematoda.* Each fluke has dendritic intestinal caecae which extend to the posterior end of the animal. The testes are large and branched and are arranged in tandem behind the ovary. The ovary is small and dendritic and is located near the acetabulum. The ovary leads to a short, coiling uterus placed between the ovary and the preacetabular cirrus pouch. A distinguishing feature of *Fasciola* is the extensive vitelline follicles which occupy much of the lateral body. A large related fluke, *Fasciola gigantica,* found in herbivores in Africa and Asia, can be 25 to 75 mm long (Feachem *et al.* 1983).

The adult liver fluke lives in the bile duct and its eggs are carried along with bile to the gut and from there they are passed out along with the faeces. The gall bladder may act as a reservoir of ova. The ovum of *Fasciola hepatica* is very large, 120 μm to 180 μm long and 60 μm to 100 μm wide, thin-shelled and ellipsoidal in shape. The ova embryonate in water in 14 to 17 days at 22°C. Hatching occurs in response to a light stimulus. An operculum opens at the tip of the ovum and the larval miracidium swims out. The life of a hatched miracidium is about 24 hours, and during this period it locates and penetrates a mollusc host, which in Europe is the snail *Limnea trunculata.* The miracidium reaches the digestive gland of the snail via the lymph channels. Each miracidium can give rise to as many as 600 cercariae via the intermediate development stages of sporocyst and redia. This process is known as polyembryony. Emergence of the cercariae occurs at around five to six weeks after the initial snail infections. On emergence from the snail, a cercaria anchors itself by means of its oral sucker to a suitable substrate such as grass, loses its tail and secretes a cyst to become an encysted metacercaria.

The metacercaria is ingested by a definitive host such as sheep and passes to the small intestine where excystment occurs, releasing an immature adult. The juvenile fluke will penetrate the intestinal wall, enter the coelom and creep over the viscera until the liver capsule is reached. The young fluke will burrow into the liver parenchyma and wander for about two months, feeding and growing until it reaches a bile duct.

The structure and permeability of the shell and vitelline membrane of the egg of *Fasciola hepatica* were studied by Wilson (1967), who hydrolysed shell material with hydrochloric acid and found, using paper chromatography, that the major amino acid components were glycine, aspartic acid, glutamic acid and serine, which implied that the protein was a fibrous type. Wilson deduced that the shell of *F. hepatica* was freely permeable to small molecules, while the vitelline membrane was the main barrier to permeability.

Acute fluke disease is caused by the damage incurred by the liver when the young flukes migrate through the liver parenchyma. An inflammatory reaction occurs within the liver tissues and this manifests itself as abdominal pain, often resulting in death due to haemorrhage and anaphylactic shock.

Black disease is a condition associated with the fluke migrations through live tissue. Damage caused by these wanderings acts as infection foci for *Clostridium novyi*. Sheep often die without any evidence of illness owing to subcutaneous haemorrhage and focal liver necrosis.

Chronic fluke disease is caused by the presence of adult flukes in the bile duct. The disease is characterized by gradual loss of weight, progressive weakness, cirrhosis, anaemia, bottle jaw oedoema and hypoproteinaemia. In cattle, chronic fluke disease is associated with thickened and calcified bile ducts.

It is recognized that the most favourable criteria for *Fasciola* ova development are a temperature above 20°C and water with a neutral pH and containing dissolved oxygen (Smyth 1976, Rawcliffe & Ollerenshaw 1960). In addition, the hatching of the fluke ovum needs a light stimulus and also the release of a hatching enzyme which attacks the opercular seal of the ovum. At temperatures below 10°C, eggs do not develop but nevertheless survive for several months, giving rise to the seasonal pattern of *Fasciola* infection in animals in temperate climates (Feachem *et al.* 1983).

The principal means of control of fascioliasis is to protect crops eaten raw, such as watercress and lettuce, from contamination by animal faeces. *Fasciola* eggs develop in dilute liquid manure, as well as water, but if animal slurry is used as fertilizer, then treatment that inactivates *Ascaris* eggs should also be effective on *Fasciola* eggs (Feachem *et al.* 1983).

5

Ascaris and other parasitic roundworms

5.1 *ASCARIS* SPECIES

The roundworms of the genus *Ascaris* are very important, because their infections are very common worldwide and because their eggs are persistent in the environment and are likely to survive conventional wastewater treatment processes. Ascariasis was estimated by the WHO to have caused over 800 million infections in Africa, Asia and Latin America in 1977 to 1978, resulting in 20 000 deaths (Warren 1988). *Ascaris* are classed as Nematoda, roundworms, and the two species of importance are the human roundworm *Ascaris lumbricoides* and the pig roundworm, *Ascaris suum*. The two species resemble each other closely, which has both advantages and disadvantages. The resemblance is advantageous, because the pig roundworms are much more readily obtainable for experimental studies. It is disadvantageous, however, because their identities become confused in environmental studies, particularly where pigs are kept in proximity to people.

Nematodes, roundworms, are circular in cross-section and cylindrical in shape, hence their name. They are in general unsegmented and have separate male and female sexes, contrasting with the hermaphroditic nature characteristic of the cestodes or tapeworms. They also possess a body cavity, and a distinct mouth, oesophagus and anus, which are important in their identification (Jeffrey & Leach 1975).

5.2 *ASCARIS LUMBRICOIDES*

The human roundworm *Ascaris lumbricoides* is the largest intestinal nematode parasitic on humans and one of the most commonly occurring human helminths. It has been estimated as infecting a thousand million people worldwide, with children particularly susceptible. The infection is not harmless, and has a fatality rate of about two cases in ten thousand (Feachem *et al.* 1983). Humans are the definitive hosts, although the eggs may be dispersed by domestic animals in their faeces.

5.2.1 Life cycle of *Ascaris lumbricoides*

The head of *Ascaris lumbricoides* is characterized by having three lips. The female roundworm is about 5 mm in diameter and 200 to 400 mm long, while the males are rather thinner and shorter. The female roundworm can produce 200 000 eggs in a day, which pass out in the faeces of the victim, but not all roundworm eggs are infective. Some eggs are infertile, and even in a fertile egg, the first-stage larva in the egg needs to develop into a second-stage larva for the egg to become infective. Infection occurs by ingestion of infective ova, on food or fingers contaminated with infected human faeces, for example. The larvae, about 25 μm long, hatch in the victim's duodenum, penetrate the intestinal mucosa and pass through the liver and heart to the lungs, where they develop further and grow to about 2 mm in length. The cycle is completed when the larvae ascend the respiratory system and are swallowed, and grow to maturity in the intestine. The adult worms then live for about a year, and typically ten or twenty worms can parasitize a host at one time. A heavy infection may involve over a hundred worms.

Infertile eggs may be produced if only female roundworms are present, or if they are excreted before being fertilized. Eggs will be passed in the faeces of the victim as long as adult female worms remain in the intestines. Fertile eggs are ovoid, about 50 to 70 μm long and 40 to 50 μm in diameter. The eggs take one or two weeks to become infective, depending on the environmental conditions into which they are passed, but can survive and remain infective for several years. Humid and warm, but not sunny, conditions are ideal for egg survival and development, with an optimum temperature of about 25 to 30°C reported.

5.2.2 Ascariasis

With mild infections of a dozen or so worms, the host may feel little more than slight discomfort. However, the presence of only a few worms is potentially dangerous, as both the adult and larval forms of *Ascaris lumbricoides* are pathogenic. Adult worms can block the intestines, bile ducts and windpipe and cause appendicitis, and migration of adult worms, carrying intestinal bacteria with them, to the liver can cause multiple abscesses on the liver and death. The worms can also consume a significant proportion of the victim's nutrient and vitamin intake, which is important if the victim is already undernourished. The larvae can cause inflammation of the lungs, coughing and difficulty in breathing.

5.2.3 Epidemiology

The occurrence and spread of ascariasis is closely associated with poor sanitation and faecal contamination of food and living space, especially indiscriminate defaecation by children. An estimated four million people were infected in the USA, according to Warren in 1974 (cited by Snowdon *et al.* 1989a) with 0.2% of pre-school children suffering intestinal obstruction. The incidence of infection is higher in women in some areas, presumably because of their contact with children and cleaning up after them, and their handling vegetables from contaminated soil. Sewage workers have an increased risk of infection (Feachem *et al.* 1983). As there is no other intermediate host involved, and eggs can remain infective for very long periods, there is an ever-present risk of re-infection.

5.3 *ASCARIS SUUM*

Female *Ascaris suum* (200 to 350 mm long) are much larger than the males (150 to 310 mm long). The males are distinguished by a curved posterior end bearing a slit-like anal opening from which extends a pair of copulatory spicules. The cuticle of *Ascaris* has transverse markings and may be brown in colour, owing to quinone-tanned proteins. The site of the excretory canals is marked by two broad brown lateral lines, and the position of the nerve cords is marked by dorsal and ventral lines which are visibly white. The female has a narrow constricted band positioned approximately one-third the distance from the anterior end, and this bears the opening of the vulva. The male also has a sub-terminal cloaca.

5.3.1 Life cycle of *Ascaris suum*

The life cycle of *Ascaris suum* parallels that of *Ascaris lumbricoides,* with a pig instead of a human host, and Feachem *et al.* (1983) state that there is evidence that *Ascaris suum* larvae may cause lung inflammation in humans and may even develop to maturity.

Ascaris suum has a direct life-cycle, and each female *Ascaris suum* has a peak daily egg production of about 2.7 million. The egg contains a single cell that can develop into an infective second-stage larva in as little as 20 days. The eggs are ingested by swine and pass to the small intestine where they hatch to release the second-stage larva. This larva then sheds its cuticular sheath, burrows through the gut wall and reaches the liver via the portal drainage. In the liver, a second moult occurs, resulting in the release of a third-stage larva. Four to six days after infection, the third-stage larvae will appear in the lungs and from here they ascend the respiratory tract and are swallowed so that the small intestine is reached in eight to ten days.

5.3.2 Ova of *Ascaris suum*

The eggs of *Ascaris suum* are round to ovoid in shape and have a thick transparent inner shell and an outer cover of an albumen-like material that gives the egg an appearance of being covered with small scales. The freshly passed egg is golden brown in colour owing to staining with bile pigments. The average size of the eggs is 65 μm by 45 μm. Some eggs may lose the albumen coat, and are known as **decorticated eggs**, allowing the structure to become visible.

In certain infections of ascariasis, only female worms are present, or the number of males may be low. As a consequence, unfertilized ova may be released. These are characterized by a long narrow shape with egg contents appearing as unorganized refractile granules.

5.3.3 Porcine ascariasis

Ascaris suum worms are pathogenic both as larvae and as adults. In response to larval presence in the liver tissue, a hypersensitivity reaction results, with allergic inflammation and eosinophilia. The inflammation heals by fibrosis, giving the milk-spot lesions that may cause the entire organ to be condemned by meat inspectors as unfit for human consumption. The inner lobular septa of the liver offer little resistance to movements of *Ascaris* larvae and the resulting economic loss due to condemned livers is an important aspect of swine ascariasis.

Larvae in the lungs cause haemorrhages followed by hyperaemia, oedoema and eosinophilic infiltration. Swine may exhibit 'thumps' due to respiratory embarrassment, with breathing being rapid, shallow and characterized by forced coughing. In severe infections, pigs may easily die.

The presence of large numbers of worms in the small intestine may result in diarrhoea, emaciation and intestinal destruction. Chronic ascariasis is distinguished by emaciation, eggs in faeces, lesions of chronic intestinal pneumonia and hepatic fibrosis. The adhesive properties of *Ascaris suum* can result in teat contamination in sows with a low reservoir of infection which results in infection of piglets before weaning. The adult worms have a tendency to wander when food is short and migrations to the stomach and bile duct are not uncommon.

Modern antihelminthics are highly efficient in treating ascariasis. The success of modern drug treatment manifests itself in the decreasing number of worm-infected pigs seen in abattoirs today. It is nevertheless preferable to avoid infection in the first place, as far as practicable, because therapeutic chemicals and/or their metabolic products may find their way into the food chain in animal products used as food or through animal slurry disposed onto land. One of the present authors has experienced the embarrassment of trying unsuccessfully to carry out bacterial fermentations in milk from cows treated with penicillin!

5.3.4 Egg development in *Ascaris suum*

The ova of *Ascaris suum* are deposited on land along with the faeces of the host. Once deposited, the ova become exposed to several environmental factors: water loss, pH, temperature, ultra-violet radiation, oxygen and organic, inorganic and other chemical components of faeces and soil.

Owing to their multilayered structure of chitin and lipid, *Ascaris suum* ova are amongst the most resistant of all helminth ova. At 22°C in water at neutral pH and containing dissolved oxygen, the ova will develop into an infective L2-larva within 14 to 20 days. *In vitro* hatching of *Ascaris* has been shown to require a temperature similar to that of homoiothermic animals, a partial pressure of carbon dioxide equivalent to 5 mole per cent, a pH near neutral and non-specific reducing conditions (Fairbairn 1961).

5.4 HATCHING OF *ASCARIS* OVA

Numerous environmental factors are known to contribute to the development and hatching of parasite ova.

Fairbairn (1961) studied the *in vitro* hatching of *Ascaris lumbricoides* eggs and concluded that the hatching stimulus involved four components: a temperature similar to that of homoiothermic animals, a partial pressure of carbon dioxide equivalent to 5 mole per cent, neutral pH and non-specific reducing conditions similar to those produced by glutathione, sodium hydrosulphite, sodium bisulphite, cysteine and sulphur dioxide. Fairbairn found that this stimulus resulted in 80 to 95% hatching in three hours. A successful hatching medium was found to result when 5 g of *Ascaris* ova were shaken in bicarbonate and 5% carbon dioxide buffer in nitrogen, with 0.04-M sodium dithionate (Rogers 1958). A further study by Rogers (1961) suggested that the hatching of ova was completed in three stages. First, the host supplies a stimulus that acts on a receptor in the infective egg. Secondly, the response to stimulation is further development, often indicated by secretion of hatching or exsheathing fluid. Finally, these fluids attack the layer of the egg shell, and the infective larva emerges.

The effect of oxygen and temperature on the embryonic development of *Ascaris suum* was investigated by Brown (1928), with the principal conclusions that lowering the temperature of a culture from 30°C to 21°C increased the development time. Eggs of *Ascaris* did not develop beyond the four or eight-cell stage at 37°C, and degenerated when maintained for some time at this temperature. The rate of oxygen consumption by *Ascaris* eggs was proportional to the rate of development of the eggs, and the amount of oxygen consumed by the eggs was the same whether the development took place at 23°C or 30°C. The amount of oxygen consumed by a single egg was very small, about 2.5 to 3.1×10^{-6} ml, although eggs at 30°C appeared to consume slightly more oxygen than eggs exposed to lower temperatures. Finally, oxygen consumption by developing eggs was found to be regular, and no one stage consumed more than any other stage. It was in fact Brown (1928) who devised the method of developmental classification for *Ascaris* eggs as: one cell, early morula 2–16 cells, late morula 16 cell to complete morula, tadpole, incurved morula to early vermiform stage, motile embryo from early vermiform stage to the completely formed embryo. The developmental classification categories for parasitc ova used in this discussion was adapted and simplified from Brown's.

The developing embryo is separated from the surrounding environmental factors by the egg shell, and it is the structure of the shell which forms a selective barrier to factors external to the egg. Bird and McClure (1976) outlined the structure of a typical nematode egg shell. They concluded that it consists of three basic layers, vitelline membrane, chitin and lipid, that it contains protein, chitin and lipid, and that its permeability is related to the absence or presence of the lipid layer. This layer consists of lipoprotein, in which the lipid component is an ascaroside in some nematodes. A study of four species of nematode ova by Bird and McClure (1976) revealed that the ova have a basic structure comprising vitelline membrane, chitin and lipid layers, but there is considerable variation in the thickness of these layers. They found that chemical analysis of the hydrolysis product of these shells indicated a high proline content of 35%.

Wharton (1979) worked on the nematode egg of *Syphacia obvelata,* an oxyurid nematode parasitic in the caecum and colon of the mouse, and found that the egg shell consisted of five layers, external uterine layer, internal uterine layer, vitelline layer, chitinous layer and lipid layer. Wharton observed that permeability of the shell was aided by discrete spaces in the uterine layers leading to pores in the external uterine layers.

Clarke and Perry (1980) observed that after eggs of *Ascaris suum* were transferred from distilled water into 0.1- to 10.4-M sodium chloride solution, the water content of the unhatched juveniles fell with increasing concentration of solute. These workers also showed that the egg shell was permeable, and that the loss of solutes from the egg fluid permitted an increase in the water content of unhatched juveniles about to undergo the onset of embryo development.

5.5 HATCHING OF OTHER PARASITE OVA

Ratcliffe (1968) obtained eggs of the parasite *Dicrocoelium lanceolatum* from bovine (cattle), ovine (sheep) and caprine (goat) gall bladders and subjected them to different physical and chemical factors, such as phosphate buffers, reducing agents, temperature, nitrogen and carbon dioxide. He found the principal *in vitro* stimulus depended on the pH and the presence of reducing conditions, and while hatching was affected by carbon dioxide, it was not stimulated by it. Ratcliffe also deduced that the hatching response of *D. lanceolatum* appeared to be similar to that of *Fasciola hepatica*.

Donnelly *et al.* (1984) investigated the influence of salinity on the ova of three species of *Schistosoma*. They found that ova of *Schistosoma mattheei, S. haematobium* and *S. mansoni* failed to hatch in salinities greater than 14.0%, while in salinities greater than 3.5%, there was a decrease in both the rate and percentage hatch of all three ova species. Salinities less than 3.5% did not significantly affect hatchability.

Egg shells of *Schistosoma mansoni* and *S. japonicum* were hydrolysed and analysed for amino acid composition (Byram and Senft 1979). It was found that the principal amino acid was glycine, while aspartic acid, lysine, serine and cysteine were often present; the authors postulated that disulphide linkages and tyrosyl bonds could contribute to the toughness of parasite egg shells.

5.6 SLUDGE TREATMENT AND *ASCARIS* OVA

Both sewage and animal wastes may contain substantial numbers of *Ascaris* and *Fasciola* ova (Feachem *et al.* 1980), so that both ascariasis and fascioliasis can therefore be initiated by land application of sewage, and animal wastes. Aiba and Sudo (1965) concluded that the problem of parasite eggs and their extinction becomes significant if purified sewage and sludge are re-used for agricultural and industrial purposes. These authors reported the incidence of *Ascaris lumbricoides* ova in Sunamachi sewage treatment plant in Japan (Table 5.1).

Table 5.1 — Eggs of *Ascaris lumbricoides* in sewage treatment (adapted from Aiba & Sudo 1965)

Stage	COD (mg/l)	Suspended solids (mg/l)	Number of eggs per litre
Raw sewage	375	604	80
Settled effluent	135	158	20
Raw sludge	—	49 290	700
Activated sludge	—	18 500	200

There is, therefore, a need to pre-sterilize sludge before land application. One of the present authors has observed that *Ascaris suum* and *Fasciola hepatica* ova were completely inhibited by temperatures between 50°C and 60°C (Kiff & Lewis-Jones 1984). In addition, organic acids at concentrations commonly found in sewage gave total inhibition of *Ascaris suum* ova and partial inhibition of *Fasciola* ova development. It is therefore likely that the diseases caused by *Ascaris suum* and *Fasciola hepatica* can be combatted, both within the hosts and in the environment in which the animals themselves live. The effects of temperature on *Ascaris* eggs is shown in Fig. 5.1.

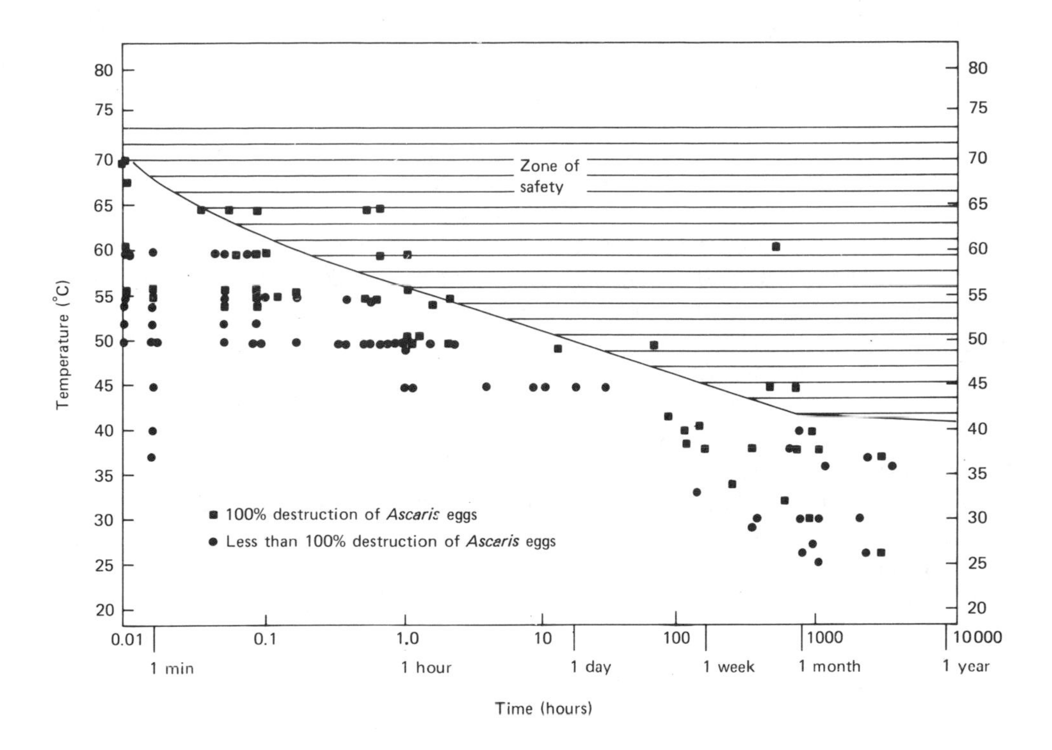

Fig. 5.1 — Effect of temperature and time on *Ascaris* eggs. From Feachem *et al.* (1983). Reproduced by kind permission of the World Bank.

5.6.1 Comparison of *Ascaris* and *Taenia*

The effects of different sludge treatments on the eggs of *Ascaris* and *Taenia saginata* were compared by Pike and Carrington (1986). Eggs of *Taenia saginata* are more resistant to temperature than those of *Ascaris* spp. at temperatures above 50°C, and complete destruction of *Taenia saginata* eggs requires heat treatment of three hours' exposure to a temperature above 60°C in pasteurization, composting or quicklime treatment. *Ascaris* ova are more resistant than those of *Taenia saginata* to lime treatment. The infectivity of *Taenia saginata* ova is significantly reduced by lime treatment at pH 12, for which at least a week's exposure is required to affect *Ascaris* viability. There may, however, be a significant difference between 'infectivity' and 'viability'.

5.6.2 Summary of effects of treatment on *Ascaris*

Some of the effects of different sludge treatments on the viability of *Ascaris* eggs reported in the literature were summarized by Pike and Carrington (1986). Some of these reports are also discussed in slightly more detail in sections 3.3.4.3, 4 and 5. Viability was estimated from the ability of the egg to embryonate on incubation and/or to produce third-stage larvae in the livers of mice.

5.6.2.1 Complete destruction of viability

Complete destruction of the viability of *Ascaris* eggs added to sludge or in sealed ampoules or 'germ carriers' was achieved by the following processes and conditions:

- *Batch pasteurization:*
  30 minutes at 60 and 65°C and five minutes at 70°C; 15 minutes at 55°C.
- *Aerobic thermophilic sludge digestion (ATSD):*
  57 hours at 49.5°C and pH 7.5; three to four days at temperatures above 53°C.
- *Reactor composting,* with sludge, sawdust and recycled compost:
  24 hours at 67° or reaching a maximum 75°C.
- *Quicklime:*
  pH 13.4, temperature 62°C in the first two hours.

5.6.2.2 Incomplete destruction of viability

Where destruction of the viability of *Ascaris* eggs was incomplete, but significant reduction of the viability of *Ascaris* eggs was achieved, the extent of viability destruction was expressed as a **survivor-fraction ratio** or survivor-fraction percentage. The survivor-fraction percentage is the ratio of the fraction of eggs recovered from sludge capable of embryonation to that in untreated eggs, expressed as a percentage. The rate of inactivation was expressed as a **specific death rate**, calculated from the survivor-fraction percentage obtained after a specified duration of treatment by assuming that destruction of viability follows first-order kinetics. Batch pasteurization was the most successful procedure of those reported, and at 53°C for one hour gave a survivor fraction ratio of 1.1% and an apparent inactivation rate of 18.75% per hour. Batch lime treatment at pH 12.5 and 29°C for just over a week resulted in a 10% survivor fraction ratio and an apparent first-order inactivation rate of just over 1% per hour. The results with anaerobic digestion at 49°C were curious. The apparent first-order inactivation rate appeared to depend on the mean retention time, and increased from 4.6% per hour with a 20-day residence time to 5.4% per hour with 16 days' retention time, and increased further with ten days' retention time. A thermodynamic parameter like a reaction rate would be expected to remain constant at a constant temperature, but the chemical composition of the sludge is also likely to change with retention time, as are the opportunities for hydraulic short-circuiting (see also section 5.6.5).

5.6.3 Pasteurization

Investigation of pasteurization conditions indicated that, although 60°C for 30 minutes should be sufficient, 65 to 70°C is needed to allow for the reduction in the effectiveness of heat treatment resulting from inhomogeneity in the system and incomplete heat penetration into aggregates of solid matter (Philipp 1981). Ova of

Ascaris lumbricoides are killed by 30 minutes at 60, 65 or 70°C (Strauch & Berg 1980). Thus, pasteurization is usually considered to consist of heat treatment at 70°C for several minutes, but the same effect can be achieved by using lower temperatures for longer times and may be more economical (Carrington 1985). Problems with re-infection have arisen where pasteurization is carried out after sludge digestion, but pasteurization followed by digestion has proved to be satisfactory. *Ascaris* ova are very heat-resistant and can be useful as convenient indicators of the effectiveness of heat treatments, particularly in respect of *Taenia* ova, which are more difficult to obtain and assay than *Ascaris* ova. As a pathogen, *Ascaris* is not an important parasite in Europe, with infections by *Ascaris* amounting to about 1300 cases a year in Britain, although as much as 50% of the population may be infected in tropical regions (Feachem *et al.* 1983).

In investigating the effects of pasteurization procedures on *Ascaris* ova (Carrington 1985), ova obtained by dissection from the lower uteri of mature female *Ascaris suum* worms obtained from an abattoir were subjected to various heat treatments and/or digestion in sludge. The viability was assessed from the proportion which developed into larvae. The viability of *Ascaris* ova was unaffected by pasteurization at temperatures below 47°C for three hours or by digestion at 35°C with a mean retention time of 13 days. Pasteurization at low temperatures (47–50°C) had little effect on viability in itself, but apparently rendered the ova sensitive to digestion. Pasteurization at higher temperatures (51–53°C) affects egg viability, but subsequent digestion had no further effect. The viability of *Ascaris* ova in sludge was, however, completely destroyed by holding at 55°C for two hours, a heat treatment likely to destroy other pathogens as well. The system must be operated so that all elements of the sludge are subjected to the heat treatment, so that a continuous system, for example, must include plug-flow characteristics.

5.6.4 Anaerobic digestion

The effects of anaerobic digestion on a range of pathogens, including ova of *Ascaris suum,* in sewage sludge have been reported (Pike *et al.* 1988). Inactivation of *Ascaris suum* ova was assessed during anaerobic digestion at 35°C and at 49°C in laboratory-scale units fed daily to give retention periods of 10, 16 and 20 days. The differences in the retention times used had little effect on the viability of recovered *Ascaris suum* ova, but the difference in the effects of the two temperatures tested was very considerable. Of the 1000 to 2000 *Ascaris suum* ova added per 100 ml, fewer than 10% were recovered, of which 63% remained viable at 35°C but only 0.6% at 49°C. Of ova stored in acidified water at 4°C, 43% lost viability.

The effects of different heat treatments with and without subsequent anaerobic digestion at 35°C on *Ascaris suum* ova were compared. Digestion alone, with a mean retention period of 13.3 days, had little effect on the viability of recovered ova. Heating only for three hours at temperatures below 50°C had little effect on viability, and heating at 50°C for two hours decreased the number of viable ova by half. Only heating for one hour at temperatures above 51°C had any real effect on the viable proportion of recovered *Ascaris suum* ova, and no viable ova were recovered from sludge heated at 55°C for 15 minutes. The viable proportion of recovered ova was considerably reduced when heat treatment was followed by anaerobic digestion, and three hours at 49°C, two hours at 50°C or one hour at 53°C, followed by anaerobic digestion at 35°C for 13.3 days, decreased the proportion of viable ova by more than 90%. It is thus clear that where a 99% decrease in viable ova is required, sludge needs heat treatment at a temperature of at least 55°C for at least 15 minutes. The digester performance differed little between heat-treated and unheated sludge in terms of gas production rate, methane content of gas and reduction in suspended solids.

Pike *et al.* (1988) pointed out that the true test of inactivation of complex organisms such as parasitic ova is their infectivity to their host. In these experiments, a proportion of the eggs added to the raw sludge were probably lysed during treatment, as they were not recovered from the heated or digested sludge, and an unexpectedly large proportion of ova autolysed during the incubation test for viability. The overall result was, nevertheless, unaffected, that thermophilic digestion or heating to at least 55°C for at least 15 minutes is required to effect 99% inactivation of *Ascaris* ova. Other conclusions were that *Ascaris* ova recovered from sludge digested at 35°C had their viability unaltered, but ova were completely inactivated by digestion at 49°C with a 10- to 20-day retention period.

5.6.5 Mathematical model of parasite inactivation

The effect of anaerobic digestion temperature and retention period on the survival of *Salmonella* and *Ascaris* ova was investigated and expressed in the form of a mathematical model (Carrington & Harman 1984a). This work aimed to quantify the appropriate procedures to prevent salmonellosis and cysticercosis, the cyst stage in cattle of the human tapeworm *Taenia saginata,* being transmitted by sewage sludge. This work investigated the effect of different retention periods and operating temperatures in laboratory-scale (4-l) anaerobic digestion on the proportion of *Salmonella duesseldorf* and ova of *Ascaris suum* surviving from a known inoculum. The ova of the pig roundworm *Ascaris suum* were used because they are more resistant to heat and ageing than *Taenia saginata* ova, and are also more readily available and easier to separate and recognize. *Salmonella duesseldorf* was used because it is easily biochemically and serologically recognizable and is not normally found in sludge.

The model tested for pathogen inactivation was

$$k_{\mathrm{D}} = \frac{1}{R\theta} \log_{10} \left(\frac{P_{\mathrm{w}} + P_{\mathrm{f}}R - P_{\mathrm{f}}R^2}{P_{\mathrm{f}}R} \right) \tag{5.1}$$

where k_{D} is the decimal pathogen decay-rate, θ the mean sludge retention period, R the fraction of the total sludge volume replaced at each feed, and P_{f} and P_{w} the pathogen content of the feed and withdrawn sludge.

Temperatures tested were 35°C and 48°C for *Salmonella* and 35°C and 49°C for *Ascaris suum* ova; retention periods tested were 10, 16 and 20 days. Inoculum levels of *Salmonella* were between 1 and 10^4/ml, and those of *Ascaris suum* ova 10 to 20/ml.

The decimal *Salmonella* decay rate was lower at both temperatures with 10 days' retention time than with 20 days', presumably due to short-circuiting effects, and tests with lower doses of *Salmonella* gave greater apparent decay rates.

The decimal *Salmonella* decay rate was about 1.4 day^{-1} at 35°C and about 3.9 day^{-1} at 48°C; determinations in a batch system gave a value of 1.6 day^{-1} at 35°C, but at 48°C, the relation did not describe *Salmonella* destruction well. The decimal decay-rate values obtained for *Ascaris suum* ova were about 0.03 day^{-1} at 35°C and about 1 day^{-1} at 49°C, although the values generally decreased with increasing retention time.

If the decimal pathogen decay-rate coefficients follow an exponential relation of the form

$$k_D(T_2)=k_D(T_1)\exp[k_T(T_2-T_1)] \tag{5.2}$$

where k_T is a temperature coefficient, then k_T is about 0.08/°C for *Salmonella* and about 0.25/°C for *Ascaris suum* ova. It has been reported that ova of *Taenia saginata* show a sharply increased sensitivity to heat at temperatures above 39°C, i.e. just above their host body temperature, and a similar sharp loss in viability has been shown to occur with *Ascaris* ova between 35° and 45°C.

5.6.6 Plant-scale anaerobic digestion

Laboratory-scale tests are essentially idealized compared with full-scale operation, because small-scale units are generally more thoroughly mixed, and thus have a closer distribution of residence times. In addition, sludge withdrawal in full-scale operation is often carried out by overflow during feeding, where the output sludge is likely to be contaminated by the raw sludge feed. Operation of a full-scale sludge pasteurization plant was described (Pike *et al.* 1988). The thickened (8% dry solids) sludge is heated to 70°C by submerged combustion of biogas from the following anaerobic digestion stage, and passes in plug-flow through a holding tank with a residence time of 30 minutes. The sludge is then cooled and digested anaerobically at 37°C with a mean retention period of 15 to 20 days. Microbiological tests showed that *Salmonella* spp. and cytopathic enteroviruses were completely destroyed. Rotaviruses were not detected in either the raw or the pasteurized sludge. A test in which *Ascaris suum* ova were added to the raw sludge feed indicated that *Ascaris suum* ova would be rendered largely non-viable. Destruction of faecal streptococci was not, however, satisfactory, with a median count of 14 000/ml in the raw sludge and 900/ml in the pasteurized sludge.

5.6.7 Earlier work on *Ascaris* in anaerobic digestion

In early work where *Ascaris* ova were seeded into laboratory digesters, after three months' anaerobic digestion, at presumably, ambient temperatures, the viability of the ova was hardly affected. After six months, 10% of the ova were viable; after a year, occasional *Ascaris* eggs were found to be viable (Cram 1943). The same study showed that hookworm ova survived anaerobic digestion for shorter times when compared with *Ascaris* ova; however, hookworm eggs did develop to larvae after sludge digestion for periods up to 64 days at 20°C, 41 days at 30°C and 36 days at room temperature. Cram found that *Entamoeba histolytica* cysts did not withstand the digestion process.

The comparative effect of both aerobic and anaerobic digestion on parasite ova has been reported (Reyes *et al.* 1963, Black *et al.* 1982). Reyes *et al.* (1963) subjected *Ascaris suum* eggs to 20 days of aerobic digestion in night soil at temperatures under

15°C and observed that none of the eggs developed further than the one-cell stage. When the temperature of aerobic digestion was raised from 15°C to 30°C, ova development resumed, with about 6% of viable ova reaching an embryonated stage. At 30°C and 40°C, 30% and 1% of samples respectively reached an embryonated stage after 20 days. Table 5.2 shows the inhibitory effect of temperature on *Ascaris suum* ova. It has also been observed that *Ascaris lumbricoides* ova were killed by a 60-minute exposure at 60°C or a 30-minute exposure at 65°C (Wiley & Westerberg 1969).

Table 5.2 — Hatchability of *Ascaris suum* eggs in night soil after exposure to different temperatures for different times (adapted from Reyes *et al.* 1963)

Temperature (°C)	Exposure time (minutes)	Hatchable eggs after exposure (as percentage of control)
45	60	97.8
	480	23.6
	1920	16.1
50	60	22.8
	90	9.4
	120	1.2
55	10	25.3
	20	16.3
	30	6.9
60	0.5	41.2
	1.0	27.3
	3.0	3.3

In aerobic digestion at temperatures above 50°C, all the eggs were destroyed within a few hours. Anaerobic digestion studies conducted by Reyes *et al.* showed that at 30°C, a 16-day detention time did not retard *Ascaris* ova development, while after 45 days' exposure at 25 to 30°C, up to 15% of eggs remained viable and healthy. At a temperature of 38°C and exposure times of 15 and 45 days, few remaining eggs were capable of development, while after 45 days' exposure to anaerobic digestion, the egg destruction was 85% at 25°C, 85% at 30°C and 92% at 38°C.

5.6.8 *Ascaris* ova in aerobic thermophilic sludge digestion (ATSD)

Clearly, mesophilic anaerobic digestion fails to achieve complete elimination of parasite eggs present in the original raw sewage (Hamer 1989). Thermophilic aerobic pretreatment provides a possible answer to this objection. Aerobic thermophilic stabilization carried out with air can give temperatures of 40 to 60°C, or with high-purity oxygen can give temperatures of over 60°C, as high as 80°C, depending on sludge solids content, retention time and aeration efficiency. Over three hours at 50°C are required to attain effective deactivation of *Ascaris suum* ova. In full-scale plants, short-circuiting results in a significant proportion of the sludge being inadequately heat-treated, and the mixing characteristics of the reactor appear to be as important as its temperature–time characteristics. In the system studied (Berg & Berman 1980), digested sludge was, however, withdrawn just *after* fresh sludge was fed in, which entails a risk of fresh, untreated sludge passing directly into the output. Studies of aerobic thermophilic digestion operating at 45 to 55°C with a retention time between 20 and 30 days, the interval between feeding and withdrawal was 12 to 24 hours, and occasional viable parasite eggs were found in the digested sludge (Kabrick & Jewell 1982).

5.6.9 Effect of composting on *Ascaris* ova

Composting data cited by Paulsrud and Eikum (1984) indicate that in a composting heap, over 80% of eggs of *Ascaris suum* survived at both the heap centre and surface in windrow composting for 16 weeks until the heaps were turned and the temperature exceeded 30°C. Data from a composting reactor indicated problems of uneven residence time, with material passing through the reactor centre in a day, while the retention time at the periphery could be several weeks, resulting in inadequate disinfection.

5.6.10 Effect of lime treatment on *Ascaris* ova

In Germany, eggs of *Ascaris, Toxocara, Trichuris, Capillaria, Oxyuris, Strongylus, Ancylostoma,* nematodes, cestodes, *Taenia* and trematodes have been found in sewage and sewage sludge (Schuh *et al.* 1985). About one-third of samples from 58 randomly selected sewage-treatment plants in Germany were found to contain helminth ova, predominantly of trichostrongylids and one of *Trichuris*. The effect of lime treatment on the eggs of *Ascaris suum* was tested in digested sludge. Unlimed digested sludge was found to have some inhibitory effect on the development of eggs of *Ascaris suum.* Digested sludge treated with lime to give an initial pH of 12.5 caused considerable damage to the eggs with a contact time of less than two months. It was therefore suggested that lime treatment of sludge to give an initial pH of at least 12.5 and storing for two months would reduce the risk in agricultural utilization of sludge arising from *Ascaris suum* to an acceptable level.

5.6.11 Effect of irradiation on *Ascaris* ova

Radiation is effective against most protozoa and most helminth eggs, although some developmental stages of *Ascaris* are resistant. A combination of temperature and radiation has been found more effective in destroying pathogens, including *Ascaris,* than either temperature or radiation alone (Sivinski 1975).

Irradiation by ^{60}Co at a dose of 300 krad had the effect on parasite ova that eggs of *Ascaris suum* remained viable but incapable of embryonation, although the resistance to irradiation depends strongly on the stage of development of the eggs (Havelaar 1984). Freshly harvested eggs had the highest resistance, which fell steadily if the eggs were left to hatch in well-aerated water at 26 to 30°C, while free embryos were highly sensitive. The effectiveness of irradiation was increased at higher temperatures. Re-contamination of irradiated sludge had not then been widely investigated, but irradiation does apparently improve the de-watering properties of the sludge. Eggs of *Ascaris suum* were found to show degenerative changes after exposure to 260 krad of gamma-radiation for 210 minutes (Wizigmann & Würsching 1974).

5.6.12 Survival of *Ascaris* in agriculture

Inactivation of *Ascaris* eggs is promoted by dry and sunny conditions leading to desiccation of the eggs. The survival of *Ascaris* eggs is favoured by moist, cool, shady conditions usually found below the surface of soil, and viable eggs thus tend to be found on root crops rather than leaf crops. In soil, the eggs survive longer than the growth period of the crops, whereas on leaf and fruit crops, the growth period of the crops exceeds the survival time of the parasites. The longest survival time recorded for *Ascaris* eggs is 14 years (Stien & Schwatzbrod 1990). Data from the literature tabulated by Stien & Schwartzbrod illustrate these effects clearly. Survival times for *Ascaris* eggs range from months in warm, dry conditions to years with cover by vegetation. The reported survival data also show an interesting comparison between *Ascaris* and *Taenia,* which survives for less than a year in soil, even in shady conditions.

Stien & Schwartzbrod (1990) investigated the survival of *Ascaris suum* eggs in three types of soil and with three vegetables, irrigated or sprayed with water seeded with eggs of the parasite. The soil types tested were sandy, clay and silt, and the vegetables were a leaf crop (lettuce, *Lactuca sativa*), a root crop (radishes, *Raphanus sativus*) and a crop with rapid regrowth (chives, *Allium schoenoprasum*). In soil, a significant number of eggs survived for three weeks, followed by a rapid decrease. The type of soil was important, and eggs survived longer in moisture-retaining soil, such as silt, than in rapid-draining sandy soil. Eggs from deep in the soil were found to retain viability longer than those on the surface, and exposure to sunlight had the expected effect on egg survival. Survival on vegetable roots depended on the type of vegetable, with *Ascaris* egg survival, rather alarmingly, much greater on radish roots than on lettuce or chives. This was attributed to the configuration of the roots. On the leaves of the vegetables, no eggs were found after ten days, which was attributed to a combination of irrigation and exposure to sunlight.

5.7 HOOKWORMS

Hookworms were estimated by WHO to have caused nearly a thousand million infections in Africa, Asia and Latin America in 1977–1978, leading to 1.5 million cases of disease and 50 000 deaths (Warren 1988). Hookworms *Ancylostoma duodenale* and *Necator americanus* cause serious problems of human infection, especially in the tropics.

5.7.1 Life-cycles of hookworms

Eggs deposited in soil hatch and release larvae which develop in three stages into an infective form which enters the human host through unbroken skin, especially the skin between the fingers or toes. *Ancylostoma duodenale* can also cause infection by being ingested in soil on unwashed raw vegetables. The larvae enter small veins or the lymphatic system and are carried to the heart and lungs of the host. The larvae enter the lung cavity and ascend the bronchial tubes and are swallowed, although

Necator americanus undergoes some further development in the lungs first. The larvae reach the small intestine, where they mature. Adult male *Ancylostoma* are about 9 mm long and 0.5 mm wide and females about 12 mm long and 0.6 mm wide, and *Necator* slightly smaller, with males about 8 mm long and females about 11 mm long (Crewe & Haddock 1985).

The sexually mature females produce eggs one or two months after the original infection. The females of *Ancylostoma duodenale* produce about 10 000 to 20 000 eggs per day, and the smaller female *Necator americanus* produces 5000 to 10 000 per day. The eggs discharged into the cavity of the small intestine are thin-shelled and develop rapidly, so that they are already partially developed by the time they are shed from the host in faeces.

5.7.2 Hookworm survival

Outside the host, in favourable warm, humid conditions out of strong sunlight, the eggs hatch in one or two days to produce larvae. The conditions favouring larval development are a light soil with suitable organic matter to provide nutrients, and temperatures of about 30°C for *Necator americanus* and 25°C for *Ancylostoma duodenale*. Larval development stops below 10°C and above 40°C. The larvae become infective third-stage larvae after about a week and can remain infective for about three months. The hookworm larvae inhabit the top 10 mm layer of soil, climb to the top of the moisture film and extend into the air. They are attracted to body heat and are activated by contact with a solid object, which is the stimulus to penetrate the skin of the host. With stronger heat, such as direct sunlight, the larvae retreat into the soil (Crewe & Haddock 1985).

Although hookworm infections are often symptomless and rarely lethal, they can cause anaemia and debility which can contribute to the death of weakened individuals and retard development in children. Hookworm eggs are thin-shelled and much less resistant to adverse conditions than *Ascaris* eggs, so that treatment methods designed to inactivate *Ascaris* should also deal with hookworm eggs (Feachem *et al.* 1983, Ketchum 1988, Jeffrey & Leach 1978).

5.8 *STRONGYLOIDES*

The nematode worm *Strongyloides stercoralis* has some similarities to the hookworms in that infection occurs by means of larvae in an infective form entering the human host through the skin, being carried to the heart and the lungs, passing up through the bronchial tubes and entering the digestive tract by being swallowed; the adult parasite then developing in the small intestine. Thenceforth, however, the life-cycle of the parasite is much more complex, and the result of infection is potentially more serious than those of the hookworms.

The adult *Strongyloides stercoralis* is very small, with mature females a couple of millimetres long which live in the mucosa of the small intestine and produce dozens of eggs daily, as a chain in a thin membrane sheath. Each egg is thick-shelled, about 50 μm×30 μm and contains a fully developed larva. The eggs hatch almost immediately to liberate non-infective larvae which migrate into the intestinal cavity, where they either pass out of the host with faeces or mature into larvae of a different threadlike form, called **filariform** larvae, which can invade the intestinal mucosa and re-infect the host. After leaving the host in the faeces, the non-infective larvae can re-infect the host by developing into infective filariform larvae on the skin of the host. They can then re-infect the host through the skin, usually in the anal region where they cause severe perianal dermatitis.

After passing out with the faeces, the non-infective larvae in dry, low-nutrient conditions, develop into the infective filariform larvae about 0.5 mm long, which infect the next host through the skin. The non-infective larvae passing out with the faeces can also develop into free-living adults in soil, where there are sufficient nutrients and moisture. The adult females can then produce eggs and non-infective larvae, which continue the free-living cycle or develop into infective filariform larvae. Infection may be symptomless, or cause diarrhoea, but in weakened hosts, the larvae may spread through the organs of the body, carrying intestinal bacteria with them, with fatal results. As *Strongyloides stercoralis* is passed in faeces as larvae, rather than environmentally hardy eggs, these are likely to be destroyed in normal sewage treatment. *Strongyloides stercoralis* occurs worldwide, but especially in areas with warm, humid climates and often together with the hookworms (Crewe & Haddock 1985, Feachem *et al.* 1983, Jeffrey & Leach 1978).

5.9 *TRICHURIS*

The whipworm *Trichuris trichiura* infects the human host by the host's ingestion of mature, infective eggs in soil on contaminated food. The eggs hatch in the intestine, and the resultant larvae pass to the coecum where they mature into male or female adults about 30 to 50 mm long. Female worms produce up to 10 000 eggs daily which pass out with faeces and mature into an infective form in soil. Infections are usually symptomless, but large numbers of worms can cause anaemia in an undernourished host (Feachem *et al.* 1983, Jeffrey & Leach 1978). The number of cases of trichuriasis in Africa, Asia and Latin America in 1977–1978 was estimated as 500 million (Warren 1988). *Trichuris* eggs can survive over three months at 25°C in clay soil with manure (cited by Stien & Schwartzbrod 1990).

5.10 *ENTEROBIUS*

The pinworm or threadworm *Enterobius* is the most widely distributed nematode infection, especially in children, and is unusual in that its eggs are not normally excreted in faeces and infection can be passed directly from person to person without the eggs needing to develop in soil. It is also more common in temperate climates than in warm climates and is relatively rare in the tropics, unlike most helminthic infections. Eggs may be present on bedclothes, clothes, toys, furniture and the fur of domestic animals in an infected household. Infection occurs when eggs are ingested as an airborne particle or from fingers; the egg hatches in the small intestine and the larvae develop into adult worms in the large intestine. As reproduction is sexual, at least two eggs must be ingested for infection to occur. Male worms are 2 to 5 mm long and female worms 8 to 13 mm long. Gravid females, containing 11 000 eggs, migrate to the anus and release their eggs, whence the cycle starts. Worms frequently migrate to the vagina in female hosts, and it is also considered possible that eggs may occasionally hatch in the anal mucosa, and the larvae migrate back into the host to cause re-infection (Crewe & Haddock 1985, Feachem *et al.* 1983, Ketchum 1988, Jeffrey & Leach 1978).

5.11 *TOXOCARA*

The ascarid worms *Toxocara canis* and *T. cati* occur throughout the world in domestic and wild canines and felines respectively. About 12% of pet dogs in the UK are estimated to be infected with *Toxocara.* Eggs passed in dog or cat faeces develop to an infective stage after about two weeks, and can persist for as long as four years. Reported data (cited by Stien & Schwartzbrod 1990) give *Toxocara* survival as eight months at 25°C and two years at 4°C in clay soil and manure.

Humans, usually children, become infected by ingesting eggs from food or fingers contaminated with faeces from an infected animal. After hatching, the larvae penetrate the intestinal wall and, over a period of several months, migrate to the liver and lungs, and may even reach the eye and cause blindness. *Toxocara* infection has been estimated to affect the sight of 150 to 200 children a year in the UK. Responsible dog-owners can reduce the risk of children losing their sight from *Toxocara* by worming their pets regularly and ensuring that dog faeces are not deposited on grass or areas to which children have access. Cats generally bury their faeces in soil where they have the opportunity, and children should be kept away from cat litter trays. *Toxocara* eggs are likely to be present in sewage, from dog faeces washed into sewers by rain from pavements, and, as ascarids, presumably have a resistance to inactivation in sewage and sludge treatment similar to that of *Ascaris* eggs.

6

Protozoan parasites

Protozoa are the simplest forms of animal life, and can be parasitic, dependent on another host organism, or free-living in water or soil. Some species feed on other organisms, such as bacteria or other protozoa, while other species assimilate nutrients by absorption in much the same way as fungi. Indeed, some species, such as the slime moulds, are classified both as protozoa and as moulds. Roughly 10 000 species of protozoa are parasitic (Ketchum 1988).

Parasitic protozoa have been defined in terms of population biology as micro-parasites, like viruses and bacteria, whose characteristics are small size, very high rates of reproduction within the host, causing a transient infection of short duration in comparison with the life-span of the host, and the tendency to induce immunity to re-infection in surviving hosts (Anderson & May 1979 cited by Warren 1988). Protozoa are responsible for a wide range of serious diseases in man and other animals. While certain species cause gastro-intestinal infections resulting in diarrhoea, other species are extremely valuable in sewage treatment processes. Consideration here is given to selected protozoa associated worldwide with gastro-intestinal disorders which are likely to be found in sewage and sewage sludge. Thus protozoa such as those causing malaria and leishmaniasis, although pathogenically extremely important, are not discussed.

6.1 PATHOGENIC PROTOZOA IN SLUDGE

The main groups of protozoa of particular importance in sewage and sludge are amoebae, ciliates, flagellates and coccidia. Gastro-intestinal infections can be caused by a few species of protozoa, amoebae, flagellates, coccidia and a ciliate. Protozoa are ingested as cysts which are resistant to the acid conditions in the stomach and pass through the upper digestive tract, and excyst into vegetative forms. The protozoa multiply in the intestine until compelled to form cysts, which pass out in the faeces. The cysts can survive in the environment outside the host for several weeks until ingested by another host (Ketchum 1988).

Ciliates are very beneficial in sewage treatment and perform a useful function by consuming dispersed non-flocculated bacteria. This reduces the content of nitrogen and suspended solids in the treated sewage effluent and improves the settling properties of the sludge. The presence of 1000 to 10 000 ciliates per millilitre indicates a healthy, well-aerated, well-settling sludge (Mara 1978). Only one ciliate, *Balantidium coli*, is known to infect humans.

Amoebae are commonly found in polluted water and in the intestines of many animals. While most are harmless, certain strains can cause amoebic dysentery, which can be very serious and even fatal. Amoebae may also themselves act as hosts for aquatic pathogenic bacteria, such as *Legionella*.

Flagellates include many species which are parasitic to plants or animals, and can cause serious diseases in humans. The most common flagellate in the human intestine is *Giardia*, which is a major cause of water-borne infection, and was in fact the first parasitic protozoon to be described.

Coccidia are all parasitic on animals and include *Toxoplasma gondii*, which causes the serious disease toxoplasmosis, which is particularly dangerous to pregnant women, and *Isospora belli*, which causes the much less serious coccidiosis. *Cryptosporidium* has only recently been recognized as a major cause of diarhoea, and is resistant to standard water-treatment processes.

6.1.1 Occurrence of protozoa

Parasites which might be transmitted to man as a result of land application of sewage sludge or animal slurries are those able to survive in an environment outside the gastro-intestinal tract for a protracted period as a stage in their life-cycle. Such are the cysts of some protozoa, which are about 30 μm in size, and millions of which may be shed daily in the faeces of even a mildly infected host. It has been estimated that if a human infected with *Giardia* sheds 2×10^8 cysts daily, a population infection level of 1% would lead to a concentration of about 10 000 cysts per litre in raw sewage (Jakubowski 1984). Often, ingestion of only a few parasites is needed to transmit infection. It must be emphasized that protozoal parasites are especially difficult to recover from samples at all, and enumeration may be further distorted by the lack of infectivity of a proportion of the cysts recovered.

Probably only a small fraction of the total number of cases of diarrhoea result from water-borne infection, but contamination of a water supply can result in many cases of illness. In the UK, there were 34 outbreaks of water-borne disease recorded between 1937 and 1986, resulting in at least six deaths out of a total of nearly 12 000 cases (Galbraith *et al.* 1987). Of 21 outbreaks due to public water supplies, 11 were due to contamination at source but did not cause any of the six deaths, and in eight of these 11 cases, water was unchlorinated or ineffectively chlorinated. The single outbreak of amoebiasis during this period was caused by a broken sewer contaminating a private bore-hole in 1950, resulting in 17 cases. Two outbreaks of cryptosporidiosis in 1983 and 1985, both in Cobham, Surrey, involved a total of over 66 cases and were attributed to a contaminated water supply. The single outbreak of water-borne giardiasis during this period was in Bristol in 1985, involved

108 cases and was presumed to be caused by the contamination of a water main during engineering work. In 1989, 9000 cases of cryptosporidiosis were reported, but it is not known how many of these were the result of water-borne infection (DoE & DoH 1990).

A survey of parasites in wastewater by the US EPA listed 46 pathogenic protozoa and helminths, although only three species of protozoa, *Balantidium coli, Entamoeba histolytica* and *Giardia lamblia*, were considered significant as human wastewater pathogens (Kowal 1982, cited in Snowdon *et al.* 1989a).

Cryptosporidium and *Giardia* are enteric dwelling protozoan parasites present in many species of animals, and zoonotic transmission to man via water can result in gastroenteritis and diarrhoea. Both *Cryptosporidium* and *Giardia* are recognized causes of diarrhoea in man, and, on a worldwide basis, *Cryptosporidium* is believed to be responsible for 0.12 to 23% of cases of diarrhoea (Fayer & Ungar 1986). Municipal drinking water sources as well as untreated surface water have been implicated in outbreaks of giardiasis and cryptosporidiosis in the USA (Stetzenbach *et al.* 1988). Both protozoan parasites naturally inhabit the intestinal tract, and the environmentally stable oocysts of *Cryptosporidium* and cysts of *Giardia* are excreted in the faeces. An investigation of natural waters showed that levels of bacterial indicator organisms, coliform and faecal coliform, do not act as reliable indicators of the presence of enteric protozoa (Rose *et al.* 1988). Detection methods for the cysts, the infective stage of the parasites, involve membrane filtration of large volumes of water, and the use of monoclonal antibodies specific to cyst and oocyst walls in a direct-smear indirect-immunofluorescent assay gave a reliable and accurate technique for assaying the cysts so recovered.

Attention is now being given to the protozoan pathogen *Cryptosporidium,* whose cysts are somewhat smaller than those of *Giardia lamblia* (Reasoner 1988). An operational parameter in chemical disinfection of water, comprising the product of disinfectant concentration and contact time, called the 'C-t value' is recognized as an important concept, but it should be applied cautiously to allow for non-ideal treatment conditions.

As with helminth parasites, the prevailing environmental conditions determine the survival of parasites outside the host's gastro-intestinal tract. Protozoan cysts can survive for long enough on plant surfaces to get into the human food chain, so that, for example, high levels of protozoan cysts can be found on fruit and vegetables irrigated or otherwise contaminated with wastewater (Snowdon *et al.* 1989a).

6.2 *CRYPTOSPORIDIUM*

Cryptosporidium is a coccidian protozoon, growing in its animal host as an obligate parasite, which has only recently been recognized as a human pathogen. *Cryptosporidium* infects man and domestic livestock, among a wide range of mammalian hosts, and is a common cause of gastroenteritis, which can be life-threatening in weakened individuals. Some outbreaks in the UK and the USA haved involved thousands of people (Casemore 1989, 1990). The presence of *Cryptosporidium* (or *Giardia*) in water has been found not to correlate with the standard indicators of faecal contamination, coliform and faecal coliform levels and turbidity (Rose *et al.* 1988). The microbiological criteria of water purity which have been in use for nearly 100 years cannot be expected to cover infection by parasites such as *Cryptosporidium*. Thus, although it is clear that transmission of the parasite occurs by means of faecally contaminated water, such water may meet current microbiological standards (Casemore 1989).

6.2.1 Epidemiology of *Cryptosporidium*

After *Cryptosporidium* was first described by Tyzzer in 1907, it was regarded as a harmless commensal organism in small animals until being associated with diarrhoea in livestock about 20 years ago. The first recognized case of cryptosporidiosis in humans was reported in 1976, and *Cryptosporidium* is now known to cause diarrhoea in otherwise healthy humans (Warrell 1990). Early reports of infection in livestock and in people associated with animals indicated that cryptosporidiosis was **zoonotic**, the term used to describe diseases transmitted naturally between animals and humans. Reports of infection in weakened individuals such as those suffering from AIDS, indicated that *Cryptosporidium* was an opportunistic organism, so that in all, cryptosporidiosis resulted from heavy levels of infection and/or weakened resistance to disease. This limited view was challenged by Casemore and Jackson in 1984, and evidence of cryptosporidial infections in towns, where zoonotic transmission would be unlikely, and of the association of *Cryptosporidium* with 'traveller's diarrhoea' and with person-to-person infection indicated that cryptosporidiosis was also a waterborne infection (Wright 1990). *Cryptosporidium* has been found in several freshwater environments (Rose *et al.* 1986, 1988). In two outbreaks of cryptosporidiosis in Cobham, Surrey, in 1983 and 1985, the distribution of cases matched that of the local water supply, which was derived from a spring, although *Cryptosporidium* was not detected in the spring, the treated water or the water-treatment system (Galbraith *et al.* 1987).

6.2.1.1 Cryptosporidiosis in humans

The first reported water-borne outbreaks of cryptosporidiosis occurred in the USA in 1984, caused by sewage contamination of a chlorinated well, and in the UK in Ayrshire in 1988, where oocysts of *Cryptosporidium* were found in treated water (Wright 1990). By 1986, over 4000 *Cryptosporidium* infections a year were reported (Galbraith *et al.* 1987), and provisional figures for 1989 have been given as nearly 8000 cases in England and Wales, over 8% of the total reports of gastro-intestinal infection (DoE & DoH 1990). An outbreak of cryptosporidiosis in Swindon and in Oxfordshire in early 1989 stimulated the then Minister for Water and Planning to set up an expert group to assess the occurrence, monitoring and public health significance of *Cryptosporidium* in water supplies and to advise on the protection of water-treatment and distribution systems. The expert group reported its findings in '*Cryptosporidium* in Water Supplies' in autumn 1990 (DoE & DoH 1990). *Cryptosporidium* and cryptosporidiosis had been previously reviewed by Fayer & Ungar in 1986.

The data for 1989 in the report provide an interesting comparison of the percentage of cryptosporidiosis cases in England and Wales with those in Scotland, which showed 1200 cases of cryptosporidiosis as amounting to 13% of identified causes of gastro-intestinal infection, a percentage about 50% higher than in England and Wales. Conversely, the proportions of *Giardia* infections showed the opposite trend, 7% in England and Wales but only 4% in Scotland, while the total proportions of cryptosporidiosis and giardiasis combined are both similar at 15% and 17% respectively. Proportions of other major causes of gastro-intestinal infection are similar in England and Wales and in Scotland, for example, *Salmonella* 26% in both, rotavirus 14% in both and *Campylobacter* 34% and 31% respectively. This may be a statistical quirk, or may indicate some association between infections by *Giardia* and *Cryptosporidium*.

Estimates of the current prevalence of cryptosporidiosis are as high as 20% worldwide, and *Cryptosporidium* is currently among the enteric pathogens most commonly reported. However, some over-reporting may occur as a result of difficulty in accurate identification of *Cryptosporidium* oocysts in samples. Cryptosporidiosis is also associated with other intestinal pathogens, and reported co-infections with *Giardia* suggest that contaminated water, fruit and vegetables are sources of infection (Casemore 1990).

6.2.1.2 Cryptosporidiosis in animals

As well as humans, a large number of domestic and wild animal species act as hosts for *Cryptosporidium*, which is thus regarded as ubiquitous, and young animals and children are the most likely to develop infections that give rise to symptoms (Wright 1990). *Cryptosporidium* infection is common in calves and lambs, although uncommon in pigs (Casemore 1990) and probably all wild and domestic animals are infected at some time while young (Gregory 1990). Several species of *Cryptosporidium* have now been recorded, but the conclusion of several authorities is that *Cryptosporidium parvum* is the species of importance in cryptosporidial infections of humans and livestock, while other species, such as *Cryptosporidium muris* are found in other mammals. Casemore (1990) warns that it is probably safer to use the term *Cryptosporidium* sp. in referring to human infections until more definitive information is available.

The significance of *Cryptosporidium* in animal infection has been summarized as being troublesome only when the animals have a low resistance and/or are subjected to a heavy infective dose of the organism (Gregory 1990), just as was found with human infection. Such circumstances arise, for example, in intensive husbandry of young animals, such as calves, where overcrowding not only increases the risk of infection but creates stress which lowers the animals' resistance. The development of intensive livestock rearing appears to be an example of an operation being modified so that conditions are changed to those suiting an organism ready to take advantage of them. This was summed up by Gregory (1990) in an aphorism to the effect that it is not the prevalence of the organism that reflects the incidence of the disease, rather the living conditions of the animals. It also implies that slurry from intensive husbandry operations is likely to be more heavily infected with *Cryptosporidium* than that from low-intensity farming. It further implies that procedures for handling, storing, treating and disposing of slurry from intensive husbandry must be designed and operated much more carefully than those for traditionally kept livestock. With water as an important means of transmission of cryptosporidial infection, the connection with animals indicates that one means of transmission is the contamination of water sources with animal slurry. This is clearly important in the handling, treatment and utilization of animal slurry, as well as sewage sludge.

6.2.2 Transmission of *Cryptosporidium*

The people most likely to become infected are those working with animals, young children in groups, such as in day-care centres, and male homosexuals, and, less unexpectedly, hospital workers and travellers in tropical countries. The most likely transmission routes are faecal–oral and zoonotic, so that contact with animals, ingesting contaminated water and person-to-person contact are the probable means of infection (Warrell 1990). The prevalence of cryptosporidiosis in children in institutions and day-care centres parallels that of giardiasis and indicates person-to-person infection (Birkhead & Vogt 1989). Washing hands after defaecating and before eating, which should be part of normal personal hygiene, becomes especially important during outbreaks of cryptosporidiosis and similar infections, not only in families, children's groups and hospitals, but also in water-treatment works.

In Third World countries, cryptosporidiosis in infants has been associated with bottle-feeding, as opposed to breast-feeding (Chasemore 1990). While this may indicate that breast-feeding confers a protective effect, it is likely that bottle-feeding acts as a source of infection when the infant feed is made up with contaminated water and the bottle and its teat are inadequately sterilized, from the experience of one of the present authors in West Africa and India. In Peru, *Cryptosporidium* was found to be one of the organisms associated with episodes of diarrhoea in infants, and an important route of infection was found to be in weaning foods, particularly as a result of improper preparation and cleaning of utensils, such as infection of sterile cans by a dirty can-opener, as well as poor domestic and personal hygiene generally (Black *et al.* 1989).

6.2.3 Symptoms of cryptosporidiosis

The symptoms of cryptosporidiosis are similar to those of giardiasis and isosporosis, usually watery, greenish and offensive diarrhoea, which may be accompanied by severe abdominal pain, cramps, fever, vomiting and other symptoms like those of influenza, but generally the symptoms last longer than in giardiasis. Unlike *Giardia* infections, however, infections with *Cryptosporidium* are not usually symptomless. While cryptosporidial diarrhoea is self-limiting in otherwise healthy people, it can be dangerous in weakened individuals, such as children suffering from malnutrition in tropical countries and AIDS patients (Warrell 1990).

6.2.4 Life-cycle of *Cryptosporidium*

Cryptosporidium is an unusual organism, and lacks several of the characteristics of other coccidial protozoa with which it is classified; its name derives from its apparent lack of a sporocyst structure. *Cryptosporidium* is an obligate parasite and its complex life-cycle is completed within a single host, so it is said to be **monoxenous**. The infective agent is an oocyst, which is very resistant to adverse environmental conditions. Oocysts can survive for several months in cool, moist soil and up to a year in clean water. This is not unusual for parasite ova and cysts, but *Cryptosporidium* oocysts also resist chemical inactivation by disinfectants commonly used in water treatment, animal husbandry and the home. Oocysts are however sensitive to freezing and to pasteurizing temperatures above about 65°C. *Cryptosporidium* oocysts are spherical and about 5 μm in diameter, with a characteristic seam, like a miniature plum.

6.2.4.1 Asexual cycle

An ingested oocyst of *Cryptosporidium* passes through to the small intestine, and the asexual part of the *Cryptosporidium* life-cycle starts as the oocyst splits along the seam with the aid of the host's digestive juices, and releases four naked, motile, banana-shaped sporozoites. The sporozoites attach to and enter the cell membrane of cells lining the small intestine, known as **epithelial** cells, but without actually entering the cytoplasm of the cell. In that pseudo-external location, the sporozoites develop into trophozoites, which undergo asexual reproduction and produce four or eight motile first-generation merozoites. The first-generation merozoites either mature directly into second-generation meronts or invade further cells to repeat the asexual cycle and produce second-generation meronts. The second-generation meronts undergo asexual reproduction to form four merozoites which differentiate into males or females.

6.2.4.2 Sexual cycle

These merozoites start the sexual cycle by entering further cells to develop into gamonts, either a small, male microgamont or a large, female macrogamont. The female macrogamonts mature into macrogametes which are fertilized by microgametes released by the male microgamont, each of which produces 16 microgametes. The fertilization occurs while the macrogamete is still within the host cell. The resultant zygotes, the fertilized macrogametes, develop into oocysts. Most of the oocysts developed have thick, double-layered walls which are the transmissable, robust form of *Cryptosporidium* containing four sporozoites. About one oocyst in five lacks the thick wall and causes autoinfection as the oocyst membrane ruptures to release its sporozoites in the intestines of the host. These sporozoites then invade further cells to continue the life-cycle within the same host. Some of the thick-walled oocysts may also release sporozoites within the host, but very large numbers of oocysts are shed in the faeces of the host into the environment, ready to infect a further host. The DoE–DoH report states that 10^{10} oocysts are excreted daily by infected calves for as long as two weeks, and a similar excretion rate would be expected in human infections.

6.2.5 *Cryptosporidium* in wastewaters

The resulting amounts of *Cryptosporidium* oocysts in wastewaters depend on the proportion of infected hosts contributing to the flow, as well as the usual sullage dilution, but 14 000 oocysts/l have been reported in raw sewage and 1300/l in activated sludge effluent, for example (Smith & Rose 1990). A level of 5500 oocysts/l was reported as an average value in raw sewage, with similar values for irrigation canals, in high-density agricultural areas and cattle pastures or areas heavily contaminated with domestic effluent, and 1400/l in chlorinated secondary effluent (Rose *et al.* 1986). The levels in surface or drinking water are usually so low that, as with estimations of protozoal cysts in general, as much as half a ton of water may have to be filtered to obtain a reasonable number of oocysts for assay. Casemore (1989, 1990) noted that the methods so far developed for recovering and identifying oocysts of *Cryptosporidium* in water are complex, time-consuming, labour-intensive and probably insensitive, and give no indication of oocyst viability.

6.2.6 Resistance of *Cryptosporidium* oocysts

6.2.6.1 Chemical disinfectants

The resistance of *Cryptosporidium* oocysts to disinfection was reviewed in the DoE–DoH report (Gregory 1990) which included a report specifically on the effect of free chlorine on the viability of *Cryptosporidium* sp. oocysts isolated from human faeces (Smith & Rose 1990). In general, chemical disinfectants at concentrations normally used in agriculture and water-treatment have little effect on *Cryptosporidium* oocysts.

6.2.6.1.1 Chlorine

Free chlorine has been found ineffective at inactivating *Cryptosporidium* oocysts in 24 hours at concentrations less than 8000 mg/l, independent of pH and temperature, and over 16 000 mg/l is needed in some circumstances. The level routinely used in water treatment is 5 mg/l or less.

6.2.6.1.2 Ozone

When chlorine is used as a disinfectant, there is a risk of formation of carcinogens such as trihalomethanes, and ozone is a promising alternative (Vaughn *et al.* 1987). Ozone has proved to be effective in inactivating 99% of oocysts at levels applicable in water treatment, using concentrations of 1 or 2 mg(ozone)/l and contact times of seven to 20 minutes (Peeters *et al.* 1989). Ozone is, however, a very reactive substance, and a large number of undesirable products can be formed when ozone is in contact with organic materials. Ozone is thus potentially most useful as a polishing treatment after the maximum amount of contaminants have been removed by other means. Ozone is probably impracticable for animal slurry or sewage sludge, and the generation of ozone at ground level is an environmental hazard.

6.2.6.1.3 *Other chemicals*

Chlorine dioxide and hydrogen peroxide have proved to be effective but only at high concentrations. Over 1000 mg/l of hydrogen peroxide is required to inactivate oocysts, although its effect was enhanced by the presence of protein, which rendered chlorine dioxide ineffective. Ammonia at 5 to 10% and formaldehyde at 10% were effective in killing *Cryptosporidium* oocysts after 18 hours' exposure.

There are other alternatives to chlorine which have not apparently been tested on *Cryptosporidium* oocysts. The use of *N*-halamine compounds as disinfectants has the advantage of stability in aqueous solution and in the solid state (Williams *et al.* 1988). The *N*-chloramine compounds 3-chloro-4,4-dimethyl-2-oxazolidinone (CDMO) and 1,3 dichloro-4,4,5,5,-tetramethyl-2-imidazolidinone (DCTMI) tend to need longer contact times than free chlorine, and while *N*-bromamines give faster disinfection, they are less stable than their *N*-chloramine analogues. Experiments with *Staphylococcus aureus* suggested that the disinfection effect of 3-chloro-4,4-dimethyl-2-oxazolidinone is due to the compound itself, rather than the small amount of free chlorine formed at equilibrium in water by hydrolysis. The disinfectant action of the *N*-bromamine analogue of this compound was due to both the compound itself and the free bromine formed at equilibrium by hydrolysis, and was a more effective disinfectant than 3-chloro-4,4-dimethyl-2-oxazolidinone.

6.2.6.2 Physical methods of disinfection

Heat treatment is the most practicable disinfection method for sludges with absolute filtration for otherwise clear water. Irradiation has aroused interest, but is so far untried.

6.2.6.2.1 Heat

Heating appears to be the most effective method of inactivating *Cryptosporidium* oocysts, for example 30 minutes at temperatures above 65°C, and moist heat at 45°C has been reported as effective after 20 minutes. Calf faeces containing *Cryptosporidium* oocysts lost infectivity after drying for up to four days.

6.2.6.2.2 Irradiation

The effects of radiation, ultraviolet, gamma- or electron-beam, have not apparently been tested. As noted earlier, electron-beam irradiation has proved to be effective in inactivating ova of *Ascaris suum* and *Schistosoma mansoni* (Levaillant & Gallien 1989) and in disinfecting chlorinated and unchlorinated secondary wastewater in respect of inactivating indicator bacteria (Waite *et al.* 1989). A relatively high proportion of *Cryptosporidium* oocysts has been found to remain in secondary effluent after sewage treatment (Rose *et al.* 1986), about 25% on average, so that irradiation is worth testing on *Cryptosporidium*.

6.2.7 *Cryptosporidium* in sludge

The effects of sewage treatment on *Cryptosporidium* oocysts were reviewed by Pike (1990). Parasite eggs generally tend to be concentrated in sludge during sewage treatment, mainly in the primary settling stage. However, *Cryptosporidium* oocysts are smaller and less dense than helminth ova, and with a diameter of 5 μm and a specific gravity of 1.08, the Stoke's Law settling velocity of *Cryptosporidium* oocysts in water at 15°C has been calculated as 3.5 mm/h, less than 1% of that of *Ascaris lumbricoides* (480 mm/h) and *Taenia saginata* (830 mm/h), and only one-fifth of that calculated for *Entamoeba histolytica* (18 mm/h). This means that *Cryptosporidium* oocysts are very unlikely to be removed from sewage in sedimentation operations unless they aggregate or are attached to much larger and/or denser particles.

From the results reported by Rose *et al.* (1986), about 75% of *Cryptosporidium* oocysts were removed from raw sewage in activated sludge treatment, including chlorination of the secondary effluent, so that both sewage sludge and treated effluent constitute contamination hazards in respect of *Cryptosporidium*. The most effective means of inactivation of *Cryptosporidium* oocysts is heat treatment, so that thermophilic digestion, aerobic or anaerobic, should be suitable for disinfecting sludge. The treated effluent is much more of a problem, because of the large volumes involved. Chlorination at practicable levels is ineffective, sand filtration is only partly effective (Rose *et al.* 1986) so that absolute filtration using a medium small enough to trap *Cryptosporidium* oocysts appears to be the only certain means of removing oocysts from treated effluent. The same problems occur in drinking water treatment, which also produces large volumes of product, which may contain low numbers of oocysts. Drinking water treatment also produces chemical and biological sludges, from flocculation and sand filtration, in which any oocysts will have been concentrated. Clearly, these should be thermophilically treated in the same way as sewage sludges.

6.2.8 *Cryptosporidium* in animal wastes

Cryptosporidium constitutes a considerable problem in dealing with animal slurry, and, by implication, sewage sludge. *Cryptosporidium* is resistant to inactivation by standard disinfectants. Animals are a major reservoir of the organism and, when infected, produce very large numbers of robust oocysts. Animals are at greater risk of infection by *Cryptosporidium* in intensive rearing, partly because of the proximity to other animals and partly because of stress from overcrowding and bullying. Intensive animal husbandry creates waste disposal problems because there may not be sufficient land available for manure disposal, which is the standard method in conventional farming. Intensive animal husbandry does, however, have the advantage that the accumulation of manure and animal slurry can be centralized, which makes slurry treatment both practicable and economic. Management of livestock effluents is clearly of great importance in controlling the dissemination of *Cryptosporidium* in the environment, and in minimizing the risk of infecting people working with the animals and of contaminating water supplies.

Treatment of manure, other than storage, before disposal is not commonly practised. Stacked manure may be turned regularly to enhance composting, to give a low-odour product suitable for horticulture. Some slurry is digested aerobically or anaerobically to reduce odour and increase stability, with the methane from anaerobic digestion providing fuel for power generation and heating animal accommodation. Clearly, thermophilic digestion should be encouraged to inactivate parasites before land spreading.

No information appears to be available on the effect of digestion on *Cryptosporidium* oocysts. However, *Cryptosporidium* is classed in the same sub-order Eimeriina as *Eimeria*, and the protozoon *Eimeria tenella* is a common enteric pathogen causing economically devastating coccidiosis in poultry. The effect of anaerobic digestion on occysts of *Eimeria tenella* has been reported (Lee & Shih 1988). Protozoan oocysts were subjected to mesophilic (35°C, 10 days) and thermophilic (50°C, five days) digestion in poultry manure, and their infectivity tested after recovery from the digested manure. The oocysts to mesophilic digestion remained moderately infective, while those in thermophilic digestion lost all their infectivity. At 35°C, however, oocysts suspended in digester liquor were more strongly inactivated than oocysts suspended in saline, indicating that temperature is not the only lethal factor involved in anaerobic digestion. Data cited by Gregory (1990) indicate that *Eimeria* oocysts in slurry are inactivated after three hours at 55°C or two weeks at 35°C. Storage at 20°C inactivated 17% of *Eimeria* oocysts in slurry after 30 days, and storage at 4°C had no effect, presumably also after 30 days.

6.2.9 Toxoplasmosis

Toxoplasmosis is caused by *Toxoplasma gondii* which belongs to the same sub-order Eimeria as *Cryptosporidium* and *Isospora belli*. *Toxoplasma gondii* is an intracellular protozoan parasite, and toxoplasmosis is one of the commonest parastitic infections of humans and livestock. Infections caused by *Toxoplasma* may be fatal (Crewe & Haddock 1985), but can also stimulate immune responses which limit subsequent infection by other protozoa and even by viruses and fungi (Nelson 1988). The domestic cat is the primary source of toxoplasmosis, but *Toxoplasma gondii* has been reported in a wide range of wild and domestic mammals and birds as well as humans.

The life-cycle of *Toxoplasma gondii* has several similarities to that of *Cryptosporidium*, involving both sexual and asexual phases. The rapidly growing asexual form, called a **tachyzoite**, enters the host cell and divides every five to seven hours. Tachyzoites are about 5 μm long and 2.5 μm wide with one end pointed. Some of the tachyzoites transform into a dormant asexual form within cysts in muscle or brain cells and probably remain in the host for life. When ingested by a cat, the encysted form differentiates in the cells lining the intestines into male microgametes and female macrogametes, which fuse to form a zygote. The zygote synthesizes a rigid, impermeable wall to become a single-celled oocyst which is excreted in the faeces. Outside the host, the oocyst becomes infectious after the single cell within the oocyst divides three times successively to form eight sporozoites.

The next host becomes infected by ingesting sporulated oocysts from food contaminated with cat faeces. The ingested oocysts hatch in the intestines to release sporozoites, which multiply intracelluarly as tachyzoites to complete the cycle. Infection can also occur when the dormant asexual encysted forms are ingested by an animal other than a cat, whereupon tachyzoites are released which multiply to cause an acute infection and transform into the dormant asexual encysted form (Pfefferkorn 1988). The sexual cycle does not occur in humans, but in pregnant women the tachyzoites can pass across the placenta and infect the foetus. This may occur in about 0.5% of pregnancies in Europe, may affect 40% of foetuses and the infection may be fatal. About 1% of cats excrete infective oocysts in their faeces, so that it is thus very important to ensure that pregnant women do not handle cat litter or otherwise come into contact with cat faeces (Crewe & Haddock 1985).

Humans can thus be infected by *Toxoplasma* in three ways, by tachyzoites in the womb, by ingesting sporulated oocysts in food contaminated with cat faeces and by ingesting the dormant encysted parasite in the muscle or brain of an infected animal. Infective cysts of *Toxoplasma* are present in a high proportion of mutton and pork products, which should be cooked at temperatures of at least 60°C for sufficient time to inactivate cysts, and hands should be washed both before and after handling raw meat (Crewe & Haddock 1985). Oocysts of *Toxoplasma gondii* are likely to be present in domestic sewage from the millions of cats kept as pets and pest control executives. The survival of *Toxoplasma* oocysts in sewage treatment is likely to parallel that of the related *Eimeria* discussed in section 6.2.8.

6.2.10 Conclusions

Heat treatment methods shown to be suitable for inactivating other parasites, such as thermophilic aerobic or anaerobic digestion, should also be effective in destroying *Cryptosporidium* oocysts, as should ionizing radiation. While such procedures are practicable for treatment of animal slurries as well as sewage sludges, heat treatment of animal slurries will be difficult to establish on a widespread basis, and animal slurries, which are predominantly disposed of onto land, constitute a major source of risk of infecting water resources with *Cryptosporidium*. Inactivation of *Cryptosporidium* oocysts in treated drinking water and treated sewage effluent is much more difficult in practice. Absolute filtration is possibly practicable, otherwise pasteurization may be possible where a water-treatment works is close to a source of cheap or waste heat, such as a power station, steelworks or a sewage treatment works where heat is generated by burning sludge or biogas from anaerobic digestion.

6.3 *GIARDIA*

In the UK, the flagellate protozoon *Giardia lamblia* is now the most common identified parasite associated with gastric illness after *Cryptosporidium*, as well as the most common brought in from overseas (Browning & Ives 1987). In 1989, 7% of identified causes of gastro-intestinal infection in England and Wales, and 4% of those in Scotland, were attributed to *Giardia* (DoE & DoH 1990). Reports of infections rose to over 5000 per year in the 1980s from about 1000 per year in the 1960s (Galbraith *et al.* 1987). Giardiasis is now considered endemic in the USA, where it is recognized as a major cause of water-borne infections (Ketchum 1988). In Africa, Asia and Latin America, there were an estimated 200 million *Giardia* infections a year in 1977 to 1978, causing half a million cases of disease, but, thankfully, very few deaths (Warren 1988). On a worldwide basis, the most commonly isolated intestinal parasites are helminths, but in countries where helminth infections are unusual, such as the UK and USA, *Giardia* is one of the most common. In specimens from humans submitted to state laboratories in the USA in 1977 and 1978, 4% were positive for *Giardia* (Snowdon *et al.* 1989a).

It should be noted that *Giardia* is known as *Lamblia* in Eastern Europe (Feachem 1983). Although *Giardia* was, in 1681, the first intestinal parasite identified, its taxonomy is still unclear. The species infecting man is known as *Giardia lamblia*, but is also referred to as *Giardia intestinalis, Giardia enterica* or *Lamblia intestinalis. Giardia lamblia* are not free-living protozoa, and the reservoir of *Giardia lamblia* is man, but *Giardia* spp., which may well be infective to man, are found in a wide range of about 40 species of wild and domestic animals, birds and amphibia.

6.3.1 Transmission of *Giardia*

Transmission occurs by ingestion of cysts of *Giardia lamblia* from contaminated water, food, body or clothes, and domestic or wild animals may also be involved; the infective dose is believed to be between 10 and 100 cysts. As with outbreaks of amoebiasis, transmission of giardiasis in developed countries is likely to arise from contamination of water mains by leaking sewage, while in developing countries, transmission is by direct ingestion of faecally contaminated food (Warhurst 1989). An intensive investigation of a significant outbreak of giardiasis in Bristol in 1985 failed to establish the source of the infection (Browning & Ives 1987). Following an outbreak of giardiasis in Vermont, USA, water-borne transmission of *Giardia lamblia* was suspected because the rates of infection were highest in people using non-municipal or unfiltered municipal water (Birkhead & Vogt 1989). The high incidence of giardiasis among children in day-care centres indicated the significance of person-to-person transmission.

6.3.2 Life-cycle of *Giardia*

On ingestion, the *Giardia* cysts resist the acid conditions in the stomach and pass to the small intestine, where they hatch to give kite-shaped trophozoites about 15 μm long and 10 μm broad, with double nuclei and four pairs of flagella. *Giardia lamblia* is a common inhabitant of the upper part of the small intestine, often with no pathogenic effect. The trophozoites multiply by longitudinal binary fission and attach themsleves to the mucosa of the jejunum, duodenum and ileum by means of an adhesive disc, and absorb nutrients from the semi-digested food of the host. They possibly may invade the lining of the bile duct and gall bladder. The vegetative trophozoites form immature or 'unripe' thin-walled pre-cysts and then the ripe, thick-walled resistant cysts that are the infective stage. While most infections are harmless, the human host can experience nausea, severe foul-smelling diarrhoea, cramps, anorexia and gas formation in the gut causing bloating and flatulence lasting several days. Nearly a thousand million cysts a day may be passed in the faeces, although the trophozoites and unripe cysts in the faeces rapidly die. The cysts remain viable for about two weeks in moist faeces, but cannot withstand desiccation (Crewe & Haddock 1985). While in water, *Giardia* cysts survive as long as 77 days at 8°C, five to 24 days at 21°C and four days or less at 37°C, although freezing and thawing cycles destroy cysts and none were to be viable after a suspension in water was brought to the boiling point (Bingham & Jarroll 1979). Cysts of *Giardia lamblia* are typically ovoid in shape and about 8 μm × 12 μm, and have been determined in raw sewage at levels of seven to 1242 cysts per litre. After allowing for the efficiency of detection and enumeration, the true level could be as much as 16 times these values. As *Giardia* are not free-living protozoa, they are unlikely to reproduce outside their host animal (Akin & Jakubowski 1986).

6.3.3 Inactivation of *Giardia*

Investigation of the susceptibility of enteric viruses and the protozoan parasite *Giardia lamblia* to disinfection and physical removal in water treatment showed that 90% removal can be achieved by complete treatment which includes coagulation, flocculation, sedimentation and 'filtration' (Reasoner 1988). Removal as high as 99% can be achieved in low-turbidity water.

Inactivation of cysts of *Giardia lamblia* with ozone was achieved within six minutes at 25°C with an ozone concentration of 0.034 mg/l, but at 5°C an ozone concentration of 0.11 mg/l was needed. Ultraviolet irradiation has been found to be ineffective in inactivating cysts of *Giardia lamblia* (cited by Akin & Jakubowski 1986).

Cysts of *Giardia lamblia* are, however, resistant to disinfection by free chlorine, especially at pH values above 8 and at low temperatures, although moderate concentrations of chlorine were effective if the temperature was high enough (Jarroll *et al.* 1981). Four factors were shown by Jarroll *et al.* to be important in the inactivation of cysts of *Giardia lamblia*, viz. chlorine concentration, water temperature, pH and contact time. Using at least 1000 cysts of *Giardia lamblia* in each trial, all cysts were inactivated at 25°C by 10 minutes' exposure to 1.5 mg/l chlorine at pH levels of 6, 7 and 8. All cysts were inactivated at 15°C by 10 minutes' exposure to 2.5 mg/l chlorine at pH 6, but at more alkaline pH levels, some survived 30 minutes' exposure. At 5°C, no cysts were inactivated by 60 minutes' exposure to 1 mg/l chlorine, while all cysts were inactivated by 60 minutes' exposure to 2 mg/l chlorine at pH 6 and 7, but not at pH 8 (Jarroll *et al.* 1981).

Certain organic *N*-halamine compounds are stable and persistent disinfectants, over a wide range of pH, temperature and water quality, for a wide variety of organisms, and an investigation (Kong *et al.* 1988) showed that 3-chloro-4,4,-dimethyl-2-oxazolidinone at low concentrations is significantly more effective in inactivating cysts of *Giardia lamblia* and *G. canis* than free chlorine.

The flagellate protozoon *Giardia muris* is pathogenic to mice and is potentially useful as a model of the human pathogen *Giardia lamblia* (Leahy *et al.* 1987). Tests of chlorine resistance indicate that *G. muris* cysts are more resistant than those of *Giardia lamblia* and are thus one of the micro-organisms most resistant to inactivation by free chlorine in existence. *G. muris* has been found to be slightly more resistant to both chlorine and ozone than *Giardia lamblia*. In comparison with other protozoan pathogens, *G. muris* cysts are about equally as sensitive to chlorine as cysts of *Entamoeba histolytica*, several times more resistant to chlorine than cysts of *Naegleria gruberi* and *N. fowleri*, but 50 times *more* sensitive than *Acanthamoeba culbertsoni* (Leahy *et al.* 1987).

The resistance to chlorine inactivation of the cysts of *G. muris* is pH-dependent, and at 25°C, chlorine was most effective at neutral or acidic pH levels. The effects of chlorine concentration and contact time generally followed Watson's Law, such that the logarithm of the disinfectant concentration is a linear function of the logarithm of the contact time required to inactivate a specified percentage of the organisms under test in that disinfectant concentration. Since the disinfectant concentration needed decreases with increasing contact time, the slope of the line is negative, and its modulus is known as the 'coefficient of dilution'. When the value of the dilution coefficient is greater than 1, the effect of dilution is more important than that of contact time, as values of the dilution coefficient greater than 1 indicate that the effectiveness of the disinfectant decreases rapidly as it is diluted. When the dilution coefficient is less than 1, then the contact time is more important. In terms of hypochlorous acid concentration, thus allowing for the effect of pH and temperature on the chlorine–water equilibrium, coefficients of dilution obtained at 25°C were 0.93, 1.55 and 1.33 at pH values of 9, 7 and 5, and 1.58 at pH 7 and 5°C. Examples of specific figures were at pH 7 and 25°C, a residual hypochlorous acid concentration of 4.1 mg/l, produced by a total chlorine concentration of 5.2 mg/l, required 7.3 minutes to achieve 99% cyst inactivation. These concentrations are well above the routinely used residual chlorine levels of 0.5 to 1 mg/l, and reduction of the temperature from 25°C to 5°C gave a tenfold decrease in the effectiveness of residual chlorine.

Chlorination is not a practicable or desirable means of disinfecting sludge, because of the massive doses of chlorine required and the concomitant risk of formation of mutagenic substances (Winkler 1984). Nevertheless, the resistance to chlorination of protozoal cysts exemplifies their resistance to inactivation in general, and emphasizes the difficulties of dealing with protozoal cysts in water treatment should they find their way into water supplies as a result of land application of inadequately treated sludge or slurry. This is particularly important because of the difficulty in determining protozoa in water, which involves the filtration of up to a tonne of water for each sample. In addition, *Giardia*, like other protozoa, do not correlate well with other standard indicators of water quality, such as turbidity, coliform or faecal coliform counts (Rose *et al.* 1988).

6.3.4 Other flagellates

An apparently non-pathogenic flagellate protozoon, *Cheilomastix mesnii*, found in the intestines follows a life-cycle similar to that of *Giardia lamblia*. Ingested cysts pass through to the colon where they hatch, multiply by binary fission and form cysts which pass out with faeces. It should, however, be borne in mind that certain organisms, notably *Cryptosporidium*, have only been recognized very recently as causes of diarrhoea. This may be owing to difficulties of taxonomy and/or identification, as with *Entamoeba histolytica* and with *Giardia lamblia*, or it may occur as a result of changes in human lifestyles, as with *Legionella*, or, as pointed out by Snowdon *et al.* (1989a), changes in habitat and movement bringing humans into contact with unaccustomed strains of organisms.

An interesting flagellate protozoon which is found in the small and large intestine is *Trichomonas hominis*, which has no cystic form. Infection is by the trophozoites from a damp environment, multiplication is by longitudinal fission and trophozoites are excreted back into the environment (Jeffery & Leach 1978). The active trophozoites can develop a resistant form without flagella which can survive at room temperature for over two weeks (Crewe & Haddock 1985). Although *Trichomonas hominis* is found in diarrhoea, this is considered insufficient evidence of pathogenicity (Jeffrey & Leach 1978). *T. hominis* is found in only a very small proportion of people in temperate climates, but may affect more than 10% of children in tropical climates. *Trichomonas vaginalis* is a sexually transmitted flagellate protozoon which is found worldwide, and infects about a million women each year in the USA. The trophozoites feed on bacteria and leucocytes in the vaginal lining and multiply by longitudinal fission. They produce a substance which injures tissue cells, generally giving rise to soreness of the vulva and vagina and a vaginal discharge, although 25 to 40% of infected women are symptomless. Some men are infected by *T. vaginalis* in the urethra, but usually without symptoms (Ketchum 1988).

6.4 *ENTAMOEBA HISTOLYTICA*

Entamoeba histolytica is a parasite of the human large intestine, and its reservoir is man, although it has been found in dogs and cats. Its infections are usually symptomless, but it can cause amoebic dysentery, consisting of bloody diarrhoea, mild fever and abdominal pain. The World Health Organization estimated that there were 400 million amoebiasis infections in Africa, Asia and Latin America per year in 1977 to 1978, resulting in 1½ million cases of disease and 30 000 deaths (Warren 1988).

6.4.1 Life-cycle of *Entamoeba histolytica*

Entamoeba histolytica enter their host by ingestion in the form of cysts, which pass unchanged through the stomach to the lower ileum. In the small intestine, the cysts give rise to metacysts, liberated from the cyst wall. The nuclei then divide and produce eight adult protozoa which emerge through an opening in the cyst wall into the colon. *Entamoeba histolytica* in its adult vegetative stage, called a **trophozoite**, multiplies in the large intestine by binary fission and encysts to form unripe cysts, known as **pre-cysts** and thence more mature or ripe cysts. Trophozoites and cysts pass out in the faeces, but the trophozoites and pre-cysts rapidly die off on exposure to air and only the ripe cysts survive in the environment outside the host, although the trophozoites are facultative aerobes (Martínez-Palomo 1988). The cysts are the infective form, and the trophozoites and pre-cysts are generally non-infective (Jeffrey & Leach 1978). Outside the body, cysts remain alive for a few days in the faeces, but, away from the faeces, the cysts can survive for long periods in suitable conditions (Crewe & Haddock 1985). An infected person can produce 15 million cysts a day in faeces. Cysts are spherical with a diameter of 10 to 15 μm and trophozoites are about 29 to 25 μm in diameter. Ingestion of a single cyst has been reported to cause an infection, but the normal infective dose is considerd to be 10 to 100 cysts (Feachem *et al.* 1983).

6.4.2 Pathogenesis of *Entamoeba histolytica*

Infections by *Entamoeba histolytica* are mostly harmless, but some involve serious enteric disease and even death. Some strains of *Entamoeba histolytica* can invade the intestinal mucosa and cause lesions, and if the entamoebae penetrate the colon, the consequences can be life-threatening. Colonization of the submucosal tissues of the colon by an invasive strain results in the formation of ulcers, from which amoebae can migrate to the liver, carrying intestinal bacteria with them, set up secondary colonization and cause a painful hepatic abscess. The lungs, brain, bladder, cervix, vagina, penis, spleen and perianal skin may also be affected. Mortality has been reported at levels ranging from 0.02 to 6% (Feachem *et al.* 1983). In the UK, amoebiasis and amoebic dysentery were rare conditions, brought in from overseas, with about 650 new infections a year in the 1980s. Only one water-borne outbreak was reported up to 1986, where a broken sewer caused contamination of the water supply from a borehole (Galbraith *et al.* 1987).

6.4.3 Identification of *Entamoeba histolytica*

Identification of *Entamoeba histolytica* can be difficult, as it is morphologically very similar to common non-pathogenic amoebae, such as *Entamoeba hartmanni* and *Entamoeba coli*. This has probably led to its prevalence being overestimated, and, indeed, *Entamoeba hartmanni* used to be called the 'small race' of *Entamoeba histolytica* until it was realized that *Entamoeba hartmanni* is non-pathogenic (Feachem 1983). The trophozoite of *Entamoeba histolytica* is elongated and shows active movement, while that of *Entamoeba coli* is more spherical and shows only sluggish movement, and the ripe cyst of *Entamoeba coli* shows a conspicuous double outline (Jeffrey & Leach 1978). The presence of these non-pathogenic amoebae is nevertheless a useful indication of faecal contamination, and thus the possible presence of pathogenic species. Incidentally, confusion can arise in literature, as the abbreviation *E. coli* is sometimes used to signify *Entamoeba coli* as well as the more common bacterium *Escherichia coli*.

6.4.4 Transmission of *Entamoeba histolytica*

The amoebae are passed from host to host by the faecal–oral route, often by raw vegetables or contaminated water, so that the prevalence of amoebiasis is directly related to local sanitation and personal hygiene in a community. Transmission routes are by polluted water, or vegetables eaten raw, such as salad plants, which have been contaminated by infected food handlers and/or the use of sewage as fertilizer. Insects may also act as carriers from sewage to food. This underlines the importance of proper sludge treatment and careful land application when sludge is used in agriculture to obviate contamination of crops and water supplies. Amoebiasis can also be sexually transmitted amongst homosexual men.

Amoebic dysentery is widely thought of as a tropical disease. In fact, it occurs throughout the world and is endemic in temperate climates as well, possibly as a result of the increasing availability of holidays in tropical climates. Although the 1950 outbreak was the only incidence of water-borne amoebiasis recorded in the UK between 1937 and 1986 (Galbraith *et al.*, 1987), the carrier rate in the UK was estimated as 2 to 5% by WHO in 1980, involving three deaths and 300 hospital cases annually. The carrier rate in the USA has been estimated as 3 to 4% but as high as 40% in male homosexuals (Feachem 1983).

The causes of outbreaks of giardiasis and amoebiasis provide a contrast between developed and developing countries (Warhurst 1989). In developing countries, transmission of protozoan pathogens is usually by direct ingestion of faecally contaminated material. Developed countries, however, are susceptible to outbreaks arising from leakage of sewage from ground water into incorrectly installed or maintained water distribution pipework. Cysts of *Giardia* and *Entamoeba* are resistant to residual free chlorine at concentrations up to 3 g/m^3 and can be removed effectively only by physical filtration. The average diameter of the cysts is about 12 μm and, without flocculation, they are not removed efficiently by rapid sand filters. The infective dose for man is little greater than one, and cysts are more likely to survive treatment of heavily contaminated water than that of relatively cyst-free water.

6.4.5 Inactivation of *Entamoeba histolytica*

The viability of the cysts of *Entamoeba histolytica* outside the host depends on the environmental temperature and humidity, and generalized figures have been given, for example, as nine to 30 days in water and up to 12 days on soil in moist cool conditions (Jeffrey & Leach 1975) and about six to eight days in soil and eight to 40 days in water (Burge & Parr 1980, cited by Snowdon *et al.* 1983), but are killed rapidly by desiccation, in as little time as 10 minutes on hands, or three days or less on vegetables in a dry climate (Feachem *et al.* 1983). Survival in sewage, sea water and soil are assumed to be the same as in water. The combined effects of temperature and time have been summarized on a log–linear plot (Fig. 6.1), giving survival times ranging from one minute at 55°C to about one month at 15°C.

The fractional inactivation rate increases fairly slowly with increasing temperature at temperatures below 37°C, but increases sharply at temperatures above 37°C. It is therefore likely that thermophilic sludge-treatment processes will be very much more effective than ambient or mesophilic treatments.

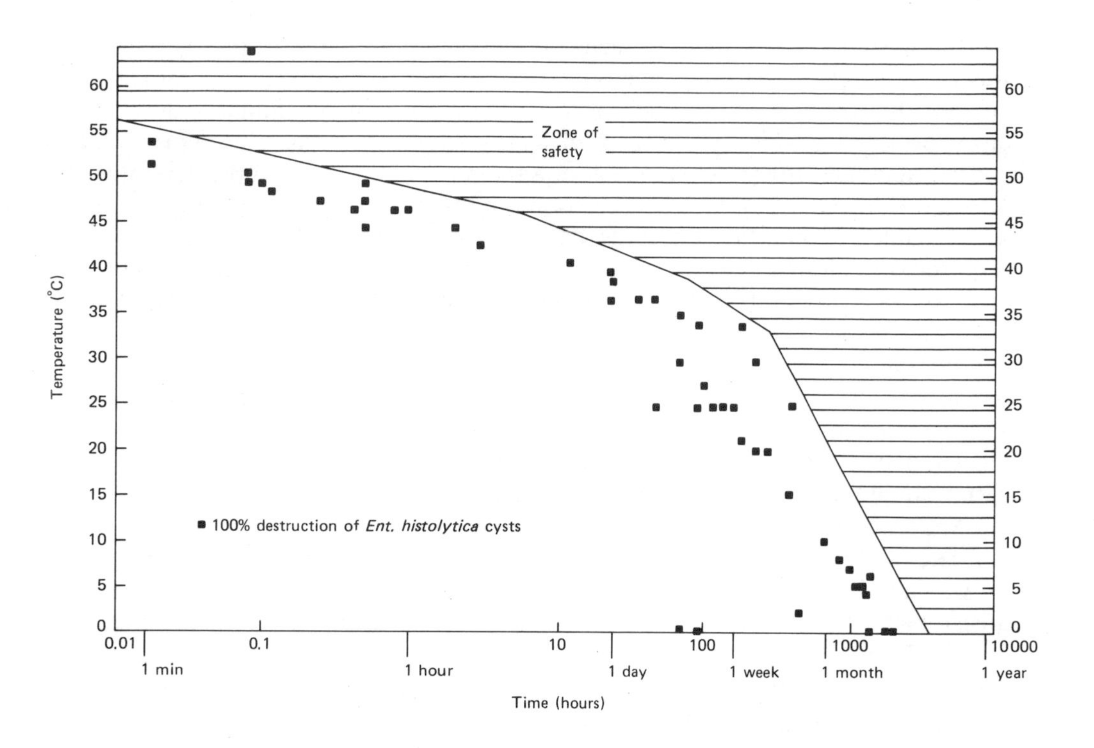

Fig. 6.1 — The influence of time and temperature on *Entamoeba histolytica* cysts. Feachem *et al.* (1983). Reproduced by kind permission of the World Bank.

6.4.6 *Entamoeba histolytica* in sewage and sludge treatment

In sewage treatment it is assumed that amoebal cysts will be separated into sludge, by sedimentation or adsorption. With a diameter of about 12 μm and a density specific gravity of 1.07, the Stoke's Law settling velocity of *Entamoeba histolytica* cysts in water at 15°C has been calculated as 18 mm/h (Pike 1990). This means that *Entamoeba histolytica* cysts are unlikely to be separated by sedimentation unless they aggregate or are attached to larger and/or denser particles. It is likely, therefore, that cysts pass through sewage treatment processes and remain in the treated effluent. Data cited by Feachem *et al.* (1983) reported 27% of *Entamoeba histolytica* cysts settling in 1.5 hours' sedimentation and 64% after two hours, and reduction of *Entamoeba coli* cysts from 52 to 27 cysts/l in primary sedimentation.

For sewage treatment processes as a whole, levels of 4 cysts/l of *Entamoeba histolytica* and 28 cysts/l *Entamoeba coli* in raw sewage were reduced to 3 cysts/l and 16 cysts/l after primary sedimentation and trickling 'filter' secondary treatment, and elsewhere, 74% and 91% of *Entamoeba histolytica* cysts were removed by two trickling 'filtration' plants, 83% by activated sludge treatment and 91% in a pilot-scale oxidation ditch (Feachem *et al.* 1983).

The fate of *Entamoeba histolytica* during anaerobic digestion was recently reported (Medhat & Stafford 1989). Anaerobic digestion is likely to be effective in reducing the levels of *Entamoeba histolytica* cells and cysts, and so help in reducing the incidence of amoebiasis in affected areas. One of the aims of the investigation was to discover if anaerobic digestion could be used to produce a pathogen-free effluent that could safely be used for irrigating crops in warm climates. Anaerobic digestion could then be especially useful in providing both biogas as an energy source and a valuable pathogen-free water supply and fertilizer for crops. The authors referred to '*Entamoeba histolytica* cells and cysts', so that 'cells' is taken to mean 'trophozoites'. The report is thus particularly interesting, first, because work reported on protozoal parasites generally deals with the effects of various treatments on cysts, and secondly, because it gives data on the rate at which *Entamoeba histolytica* consumes bacterial cells.

Entamoeba histolytica cells in saline were found to decay at a rate of about 1% per hour at 55°C, but decayed at about 25% per hour in an anaerobic digester at 55°C with a mean hydraulic retention time of 10 days. Thus thermophilic anaerobic digestion at 55°C removed 90% of added *Entamoeba histolytica* cells in eight hours, with virtually complete elimination after 12 hours.

Anaerobic digestion at 37°C reduced the cell numbers of *Entamoeba histolytica* at a rate of about 10% per hour, with some cells still surviving after 120 hours, from initial values of over one million per millilitre. The conditions in an anaerobic digester at 37°C roughly simulate those in the human gut. When the content of volatile fatty acid (VFA) was increased by the addition of magnesium acetate, the cell reduction rate increased evenly with VFA concentration to about 30% per hour with a VFA concentration of 2800 mg/l. Biogas production was lost at VFA concentrations of 3500 mg/l. Increasing the VFA can thus be used to increase the cell reduction rate of *Entamoeba histolytica*, but entails the risk of the breakdown of gas production. Measurements of the numbers of *Escherichia coli* bacteria indicated that each *Entamoeba histolytica* cell consumed 21 *Escherichia coli* cells and ten granules of rice starch in 48 hours.

6.4.7 Other amoebae

In addition to *Entamoeba coli* and *Entamoeba hartmanni* other non-pathogenic intestinal amoebae are *Endolix nana*, which feeds on bacteria and faecal fragments, *Iodamoeba butschlii* and *Dientamoeba fragilis* (Jeffrey & Leach 1978), although *D. fragilis* is probably the amoeboid stage of a flagellate, rather than a true amoeba (Crewe & Haddock 1985). Some free-living amoebae can invade mammalian tissue and cause infections of the eyes, lungs, sinuses, ears and central nervous system in humans (Crewe & Haddock 1985). A rare amoebic disease is amoebic meningoencephalitis, of which there were six cases in the UK between 1937 and 1986, caused by *Naegleria fowleri* (Galbraith *et al.* 1987) and over 120 cases reported worldwide between 1965 and 1985 (Crewe & Haddock 1985). The disease is associated with children and young people swimming in fresh water, and is usually fatal (Ketchum 1988).

6.4.8 Amoebae and *Legionella*

Naegleria and *Acanthamoeba* amoebae have been shown to act as hosts for the pathogenic bacterium *Legionella*, the causative organism of Legionnaire's disease and Pontiac fever (Harf & Monteil 1989). *Legionella* have identical aquatic habitats to those of free-living amoebae, which are therefore likely to be natural hosts for the bacterium. Culture and isolation of *Legionella* from natural waters is difficult, even though they are commonly present in the environment. In laboratory tests, Harf and Monteil showed not only that amoebae feed on *Legionella* in both suspended and surface culture, but both survived several subculturings of the amoebae. Amoebae were isolated from water and sediments in three rivers in Alsace and cysts or trophozoites subcultured. After thermal and ultrasonic lysis, the lysates were shown to be positive for *Legionella pneumophila*. In addition to demonstrating a vector for the bacterium and a convenient laboratory ecological model, it also partly explains why it is difficult to remove *Legionella* bacteria from air-condition cooling systems. Protozoal cysts in general are resistant to adverse conditions, including chlorination, so that it is likely that *Legionella* and other bacteria ingested by protozoa are protected from disinfection treatments designed to deal only with vegetative organisms. From other results reported by Harf and Monteil, it is also likely that heat treatment actually liberates bacteria from protozoa. This is analogous to problems encountered in the early days of pasteurization, when it was found that bacterial counts in milk were often higher after pasteurization than before, as a result of spore germination stimulated by thermal shock. Procedures for disinfection of air-conditioning cooling water must therefore be designed to allow for these effects. The presence of *Legionella* in an aquatic protozoon *Acanthamoeba palestinensis* in Israel suggest a mechanism that enables the bacterium to survive the winter and resist attack by biocidal agents (Shuval *et al.* 1988).

6.5 *BALANTIDIUM*

Balantidium coli is a ciliate protozoon, and is the only parasitic ciliate known to infect humans. Pigs and rats are the main reservoirs and balantidiasis is associated with communities that live in close proximity to pigs and the associated rats. In the UK, 80% of pigs are believed to be infected with *Balantidium coli*. This serves to emphasize the importance of proper disinfection of animal slurry, as well as sewage sludge, before its application to land, because of the risk of infective cysts finding their way into watercourses.

6.5.1 Life-cycle of *Balantidium coli*

Infection occurs by ingesting cysts from food, water or utensils contaminated with faeces from an infected carrier, a person, a pig or, presumably, a rat. The cysts pass unchanged through the upper digestive tract and the ciliate trophozoites excyst in the intestine. The trophozoites multiply by binary fission and form cysts which pass out in the faeces with the trophozoites. *Balantidium* differs from *Entamoeba histolytica* in that trophozoites can survive outside the host and may also be infective. Evidence that trophozoites can survive the conditions of low pH in the stomach on ingestion is, however, conflicting (Jeffery & Leach 1978, Faechem *et al.* 1983, Ketchum 1988). Cysts can survive in moist conditions for several weeks, but are rapidly killed by desiccation (Crewe & Haddock 1985).

6.5.2 Pathogenesis of balantidiasis

As with amoebiasis, most infections are symptomless, and *Balantidium* lives harmlessly as a commensal organism in the colon. On occasion, however, the organism invades the colonic mucosa and causes balantidial dysentery, producing wide-mouthed ulcers, somewhat like those in amoebiasis, and bloody diarrhoea. Infection is localized to the intestine, however, but although *Balantidium* does not spread beyond the intestines, it may cause acute appendicitis. Death has been reported as occurring in 5 to 35% of cases in the tropics, and may occur as a result of the ulceration, haemorrhage and dehydration.

6.5.3 Inactivation of *Balantidium*

Little appears to be known about the susceptibility of *Balantidium coli* to inactivation in the environment and in sewage and sludge treatment, but it is presumed to be similar to that of *Entamoeba histolytica*. Incidentally, some confusion of nomenclature may occur with *Balantidium coli* if it is abbreviated to *B. coli*, because that is also found in older literature as the abbreviation for *Bacillus coli*, the former name for *Escherichia coli*

6.6 CONCLUSIONS

As with *Cryptosporidium*, it appears that thermophilic treatments, aerobic or anaerobic, carried out according to the recommended guidelines, can give adequate disinfection of sludge in respect of protozoal parasites. There will, however, be problems in ensuring that animal slurries are properly treated. Chemical disinfection is not a practical proposition for inactivating protozoal parasites, except those in which pasteurizing temperatures or very high pH values are achieved for sufficient time.

7

Bacterial pathogens in sludge

Faeces excreted by a healthy person contain many species of commensal bacteria in large numbers, usually between ten million and ten thousand million per gram, depending on the diet of the individual. Among the most numerous and most commonly occurring bacteria in healthy people are enterobacteria, enterococci and *Clostridium perfringens*, which are frequently used as indicators of faecal contamination, and some of which can be pathogenic.

An individual infected with a pathogen causing gastrointestinal disorders excretes large numbers of the organism in faeces. These are therefore present in sewage and in certain types of sewage sludge, and constitute health hazards and other problems in sludge handling. In sludge treatment, however, bacteria might appear to constitute less of a problem than helminth or protozoal parasites. Parasite ova and cysts are very hardy and resistant to inactivation, and sludge treatments which are effective in destroying these should also be more than adequate for inactivating bacterial pathogens. However, bacterial pathogens present three difficulties not found with helminth and protozoal parasites.

Firstly, bacterial pathogens, when present in sludge, are usually present in enormous numbers. If it is at all possible for an organism to survive a particular treatment, then the larger the number of organisms present at the start of the treatment, then the larger the number of organisms likely to survive the treatment.

Secondly, bacteria are generally capable of survival and growth outside their animal host, in suitable conditions, and sewage sludge provides a very good growth medium for many bacterial pathogens. As fresh faecal material is continuously entering a sewage-treatment works, treated sludge can be rapidly re-infected with bacterial pathogens. These can then proliferate more rapidly than in undisinfected sludge, because of the absence of competition from other bacteria.

Thirdly, certain bacteria, including the extremely dangerously pathogenic *Clostridia*, can form spores which are highly resistant to adverse conditions (section 7.1.3), and are probably more resistant to sludge treatments than even helminth ova.

7.1 INDICATOR BACTERIA

7.1.1 Enterobacteria

The enterobacteria include mostly *Escherichia coli* and bacteria resembling them, known as 'coliform' bacteria. Coliform bacteria are all Gram-negative rods 2 to 5 μm long and about 0.5 μm in diameter, and are classed as **faecal coliforms** or **non-faecal coliforms**. Faecal coliforms, such as *Escherichia coli*, have the intestines of humans and warm-blooded animals as their natural and exclusive habitat. Non-faecal coliforms, such as *Klebsiella*, *Citrobacter* and *Enterobacter aerogenes*, are found naturally in unpolluted soil as well as the intestines of animals. The total number of both faecal and non-faecal coliforms, known as **total coliforms** is commonly used as a criterion of bacteriological quality for water and treated sewage effluent. *Escherichia coli* comprise over 90% of the total coliform bacteria in the fresh faeces of warm-blooded animals.

Coliform bacteria are useful as indicators because their metabolic characteristics make them readily identifiable, and because in temperate climates, they survive only a few hours or days outside their host, so that their presence indicates recent faecal contamination. The presence of faecal coliforms in a medium is a clear indication of faecal contamination, while the presence of non-faecal coliforms indicates a likelihood of faecal contamination. The absence of faecal coliforms does not prove that faecal contamination has not occurred, and in hot climates, some sewage may not contain faecal coliforms. This is because high ambient temperatures enable non-faecal coliforms to grow, but not faecal coliforms and pathogens (Mara 1978). Doubts have been raised as to the reliability of bacteria as faecal indicators, as other pathogenic constituents of faeces, such as viruses and parasite ova or cysts, can be found in natural waters free of bacterial faecal indicators. Enteric bacteria illuminated with visible light tend to change to a dormant, non-culturable state, in which substrates are not incorporated and biosynthetic processes are inactivated (Barcina *et al.* 1990). These effects appear to be enhanced in seawater, although seawater has been found to have a greater effect on *Escherichia coli* than on *Enterococcus faecalis*.

Certain strains of *Escherichia coli* can be pathogenic, and these are described in more detail further on (section 7.9).

7.1.2 Faecal streptococci

Faecal streptococci are Gram-positive, roughly spherical bacteria about 1 μm in diameter, occurring in short chains. *Streptococcus* spp. are found mostly, but not exclusively, in the intestines of humans and warm-blooded animals, but some species are found in unpolluted as well as polluted environments. Although faecal streptococci are thus not ideal as indicators of faecal contamination, they are relatively easy to enumerate, survive longer and less likely to grow than faecal coliforms. The ratio of faecal coliforms of faecal streptococci was at one time thought to provide a means of differentiating between human and animal faecal contamination. This is now considered unreliable in view of the different growth-rates and geographical distributions of the two types of organisms. However, a reliable test for differentiating between human and animal contamination would be extremely valuable to identify and remedy sources of pollution.

7.1.3 *Clostridium perfringens*

Clostridium perfringens, also known as *C. welchii*, is invariably present in faeces, and so is used as an indicator organism for faecal pollution. One of its key characteristics is that it forms spores, which survive severely adverse conditions, and so persists in the environment. This is both an advantage and a disadvantage in using *Clostridium perfringens* as an indicator organism. The presistence of the spores ensures that traces of faecal pollution do not disappear before being detected, but at the same time, it also leads to the establishment of a dormant population which may confuse the detection of subsequent pollution incidents. *Clostridium perfringens* is a Gram-positive rod about 5 μm long and 1 or 2 μm in diameter, and is an obligate anaerobe.

The resistance of bacterial spores to adverse conditions is illustrated by the results of an evaluation of the sporicidal activity of ozone, as a possible alternative to 2% glutaraldehyde or a lengthy (over 30 minutes) exposure to ethylene oxide (Rickloff 1987). Spores of both *Bacillus subtilis* and *Clostridium sporogenes* were inactivated by ten minutes' exposure to ozone-saturated water (about 10 mg ozone/l) at ambient temperature. The resistance of the spores increased after drying, with viable spores recovered after 40 minutes' exposure at ambient temperature, and no improvement was gained by increasing the temperature to 60°C. It was concluded that ozone penetration and organic loading were key factors in the disinfection effect.

Clostridium perfringens is itself a pathogen as well as a faecal indicator. It can infect an open wound and cause gas gangrene, and can cause food poisoning. The mechanism of *Clostridium perfringens* in food poisoning is discussed in more detail further on (section 7.10), as is that of *Clostridium botulinum* (section 7.11).

7.2 BACTERIA IN SLUDGES

7.2.1 Primary sludge

Preliminary treatment of sewage is aimed at removing solid material large enough and dense enough to cause damage in subsequent treatment operations, and effects virtually no removal of micro-organisms. Primary sedimentation, however, effects the settling of mainly faecal solids, so that a large proportion, 50 to 90%, of faecal organisms, including pathogens, is carried into primary sludge. Primary sludge will thus contain a bacterial population similar to that of the original faecal matter in the incoming sewage. Co-settled primary sludge will contain a lower proportion of faecal organisms than standard primary sludge corresponding to the lower number of faecal organisms in secondary sludges. The supernatant 'primary effluent' from primary sedimentation carries the residual unsettled dispersed organisms into the secondary treatment stage.

7.2.2 Activated sludge

The organisms in activated sludge are predominantly bacteria, aggregated into sludge flocs which form the ecological unit of activated sludge. Each floc is a cluster of several million bacterial cells, most of them dead or moribund, with a few per cent of living bacteria, inert organic and inorganic matter and other organisms such as protozoa and fungi. A healthy sludge contains many different bacterial species and a low diversity of species indicates faulty or potentially faulty plant operation. The bacteria in secondary sewage treatment derive from two sources, the faecal matter in the sewage and the infiltrating water, which contains aquatic bacteria and those washed in from soil.

The bacteria in activated sludge are, in practice those of aquatic origin, rather than faecal, as the temperature of the system in temperate climates favours aquatic organisms rather than faecal organisms, whose natural habitat is warm-blooded animals. The bacteria are also of aquatic origin rather than soil origin, as those in soil are predominantly Gram-positive, whereas the bacteria in activated sludge are predominantly Gram-negative, such as *Pseudomonas*, *Achromobacter*, *Flavobacterium*, *Nocardia*, *Mycobacterium* and *Bdellovibrio*. Systems which are sufficiently low-loaded to effect nitrification will also include the nitrifying bacteria *Nitrosomonas* and *Nitrobacter*. Bacteria, particularly dispersed bacteria, are consumed by protozoa grazing on the sludge flocs.

The activated sludge process provides a hostile environment for faecal organisms, and can be expected to remove a considerable proportion of faecal organisms, including pathogens, from treated effluent by starvation, predation and adsorption on to settled sludge flocs. The proportion remaining in the treated effluent and in the settled sludge depends mainly on the process operating conditions. A low-rate process with a long contact time between sludge and sewage will remove a very high proportion of faecal bacteria, as much as 99%, and a high proportion of the bacteria adsorbed on to sludge flocs will be consumed during the concomitant long residence time of the sludge. Conversely, a high-rate process with a short contact time between sewage and sludge and a short sludge residence time may remove only low proportions, with the settled sludge containing still-viable faecal organisms.

In considering the infection risks associated with sludge, the pathogen content of activated sludge is in practice almost irrelevant, as surplus activated sludge is usually thickened and mixed with primary sludge for further treatment or disposal, thus neatly re-infecting it with faecal organisms. The lowest infection risk is associated with sludge from a low-rate or extended aeration system, preferably one treating whole comminuted sewage, where no primary sludge is produced. This is unlikely to become normal practice, as it requires roughly double the aeration capacity of a plant treating primary-settled sewage. Nevertheless, the costs of prolonged secondary treatment are counteracted by the resultant savings in the costs of sludge treatment and disposal, which are likely to rise as disposal options become increasingly constrained. One of the problems of sludge treatment is that disinfected sludge rapidly becomes re-infected with pathogens, because sludge treatment operations cannot practicably be run to the same standards of asepsis as antibiotic production plants. The risk of infection is considerably reduced if only low-rate secondary sludge is being produced.

7.2.3 Percolator humus

The humus from percolator (or 'biofilter') secondary sewage treatment units is a micro-ecological system of aerobic, anaerobic and facultative bacteria, as well as fungi, protozoa and macro-fauna. Rather ironically, as perolators were sometimes known as 'bacteria beds', fungi are more common in percolator humus than in activated sludge. Bacteria are predominant in units treating domestic sewage, but the ecological balance in the humus depends not only on the nature of the wastewater being treated, but also on the position within the percolator packing. This is because percolator operation approximates to a plug-flow system, so that the wastewater changes in composition as it trickles through the bed, and affects the relative growth rates of the organisms comprising the microbiological film. Bacterial species commonly found in humus are *Achromobacter*, *Flavobacterium*, *Pseudomonas* and *Alcaligenes*, and filamentous genera such as *Beggiatoa* and *Sphaerotilus*. Fungi with an ability to attach to surfaces, such as *Fusarium* and *Geotrichum*, are naturally favoured in attached-growth systems, and other important fungal genera are *Sepedonium*, *Ascoidea* and *Subbaromyces*.

Percolator beds constitute a hostile environment for faecal bacteria, and, as with activated sludge, the removal efficiency depends on the process loading. This affects the mean residence time of the wastewater in the bed as well as the turnover of organisms in the percolator microflora. The contact time in percolators is much less than that in the activated sludge process, about an hour, compared with half a day in conventional activated sludge treatment, and percolators are conventionally operated at very low loadings, so that the proportion of faecal bacteria removed is comparable with that in the activated sludge process.

7.2.4 Treated sewage effluent

The supernatant from the clarification stage following secondary treatment is the treated sewage effluent, which is by no means sterile. It contains dispersed, unflocculated bacteria and sludge particles too small to settle in the time available in the clarifier. This is generally expressed in terms of suspended solids, turbidity or organic nitrogen content, rather than actual numbers of organisms, but may contain 1% (0.1 to 10%) of the pathogenic bacteria originally in the incoming sewage. These may be removed by tertiary treatments, similar to those used in drinking-water treatment, which will produce biological and/or physico-chemical sludges requiring disposal.

7.2.5 Genetically engineered bacteria

The transfer of genetic material between bacteria is a natural process which can occur in wastewater and other growth systems, and this ability of plasmid DNA to be transmitted between bacteria and to recombine with the DNA of the recipient organisms could impose a limitation on applications of recombinant organisms. Genetically engineered organisms may be released into the environment for agricultural purposes, and with the increasing utilization of genetically engineered organisms in industry, there is a risk that some will find their way into waste treatment systems. Deliberately released organisms can be modified to survive only until they have performed their required function, but there is nevertheless the possibility that genetically engineered DNA sequences can be transferred by bacterial conjugation to other wastewater organisms.

The transfer of non-bacterial genetic material has been demonstrated in a laboratory-scale waste-treatment unit (Mancini *et al.* 1987). Inter-species transfers of conjugative plasmid material in a wastewater treatment plant have been reported, from a laboratory strain of *Escherichia coli* to a strain of *Escherichia coli* and *Enterobacter cloacae* isolated from raw wastewater, and between different *Salmonella* spp. In the research context, plasmids from *Escherichia coli* have been transferred to such diverse organisms as *Agrobacterium tumifaciens*, *Anabaena* spp. and *Rhodopseudomonas sphaeroides* (Gealt 1988).

An investigation of recombinant DNA plasmid transmission to indigenous organisms during waste treatment has been reported by Gealt (1988). The gene transfer process has been demonstrated in a laboratory-scale (20-l) waste treatment unit, using *Esherichia coli* HB101, and was found to depend on the presence of bacteria containing conjugative plasmids. Such bacteria are commonly found in wastewater, and the frequency of conjugal gene transfer was not affected by the nutrient composition, concentration or flow rate. Experiments with various salts and detergents showed that the non-ionic detergent Triton X-100 had no effect on conjugation, the 'non-heavy' metal ions magnesium, calcium, sodium and potassium at moderately high concentrations decreased the number of transconjugants, while the anionic detergent SDS and the heavy metals zinc and iron markedly decreased the number of transconjugants at very low concentrations. It appeared that the efficiency of plasmid mobilization depended on the concentration of donor cells, as the presence of transconjugants was noted only when the concentration of donor cells in the sludge exceeded 10^7/ml.

The fate of genetically engineered micro-organisms in freshwater was investigated by Pickup *et al.* (1989), who found that genetically engineered strains of *Pseudomonas putida*, *Ps. fluorescens*, *Serratia rubidaea* and *Escherichia coli* inoculated into sterile fresh lake-water survived and maintained concentrations of 1000 to 10^5 cells/ml over 60 days. Plasmid retention ability varied between and within different genera, and their auxotrophic test organisms were undetectable after 30 days.

7.3 PATHOGENIC BACTERIA

Genera of pathogenic bacteria found in faeces include *Brucella*, *Campylobacter*, *Clostridium*, *Klebsiella*, *Mycobacterium*, *Pseudomonas*, *Salmonella*, *Shigella*, *Staphylococcus* and *Yersinia*. Over 40 different species of pathogenic bacteria have been enumerated from sewage sludge (Dudley *et al.* 1980), and animals are the principal reservoirs of bacteria that are pathogenic to man. There is an enormous and diverse bacterial population in animal slurries, with, for example, 7×10^6 to 3×10^{11} colony-forming units (cfus) per millilitre reported in pig effluent (Strauch 1978 cited by Snowdon *et al.* 1989a), although bacteria specific to livestock cannot, by definition, infect humans and vice-versa. Grazing animals are not easily infected unless they are subjected to prolonged exposure to heavy contamination or are otherwise unusually susceptible to infection for some other reason (Jones 1977 cited by Snowdon *et al.* 1989a).

Clearly, a wide range of human pathogens find their way into sewage from faeces as well as other, non-faecal human wastes. Those considered here are key bacteria causing gastro-intestinal disease and associated with transmission by faecal contamination that could occur in the handling and disposal of sewage sludges.

7.3.1 Mechanisms of pathogenesis

All the bacterial pathogens considered here can cause infections which are symptomless. Bacteria cause disease by invading tissue in a host, or producing toxic substances, or often both. Invasion of tissue can damage the tissue and stimulate inflammatory or other adverse immune reactions in the host. Toxins can damage cells or tissues or disrupt their metabolic function, and can also stimulate immune responses. Toxins produced by bacteria are of two types. **Exotoxins** are excreted by the bacteria into the surrounding medium, which may be a food material or the tissue of the host. Toxins may enter cells in the host and disrupt their function, or may be enzymes which catalyse the breakdown of the cell or tissue structure. Certain strains of both Gram-positive and Gram-negative bacteria can produce exotoxins. Exotoxins may be produced within the host, by the colonizing bacteria, or be produced outside the host, in a food material for example, and ingested. Exotoxins that act on the intestines are known as *Enterotoxins*. **Endotoxins** are associated only with Gram-negative bacteria, as they are components of the outer membrane of the cell wall characteristic of Gram-negative bacteria. Endotoxins can cause diarrhoea and stimulate the release of a pyrogen from white blood cells, causing fever (Ketchum 1988).

7.4 BACTERIAL SURVIVAL

Survival times of human-associated bacteria outside the enteric environment are shortened by adverse temperatures, low nutrient levels an competition from other microflora.

7.4.1 Soil–wastewater systems

A wide range of factors are known to affect the survival of bacteria in soil or on pasture, including the initial number of organisms, frost, moisture, pH and concentration of inorganic salts (ionic strength), permeability and particle size of soil, as well as temperature, utilizable organic material and interactions with other micro-organisms, both inhibitory and beneficial. This leads to a correspondingly wide range in the bacterial survival rates the bacterial survival rates reported (Snowdon *et al.* 1989a). The mean half-life of over 120 bacterial species was 76 hours, corresponding to k_1=0.22 day^{-1} where k_1 is a first-order reaction-rate coefficient, or roughly 1% dying off per hour (Reddy *et al.* 1981 cited by Snowdon *et al.* 1989a). The death rate approximately doubles for every 10°C rise in temperature in a temperature range 5 to 30°C.

7.4.2 Seawater

The survival of indicator bacteria and pathogens has been of particular interest in respect of standards for water used for bathing and for shellfish growing. It is now well-established that bathing in water contaminated with sewage carries a definite risk of contracting gastro-intestinal infections, probably of viral origin, and that the incidence of infection correlates best with the concentration of *Enterococcus* in the water. Assaying bacteria is now complicated by evidence that some Gram-negative bacteria can enter a dormant state in which they are viable but not culturable (Evison 1988).

Comparative studies on the survival of indicator organisms and pathogens in fresh water and seawater were reported by Evison (1988). A range of organisms was tested for survival at different temperatures, in the light and the dark, in fresh water and seawater at different salinities and concentrations of nutrients in the form of sterile raw sewage. The organisms tested were *Escherichia coli*, faecal streptococci and maroon faecal streptococci from natural populations in raw sewage, strains of *Salmonella typhimurium* types 12, 12A and 110, *Salmonella anatum*, *Shigella flexneri*, *Sh. sonnei*, *Yersinia enterocolitica* and *Campylobacter fetus* isolated from faeces, and two F-specific RNA coliphages, MS2 and f2 (section 8.3).

Results showed that, in terms of the time taken for 90% inactivation, organisms generally survived better in fresh water in the light, but better in seawater in the dark. The *Salmonella* spp. survived very much better than any other organism, the *Escherichia coli* worst, the streptococci a little better then the *Escherichia coli.* Linear relationships were found between the 90% inactivation time (T_{90}) and light intensity and between log T_{90} and temperature. The effects of salinity and nutrient concentration on organisms survival were non-linear.

The results showed that light was of overriding importance in accelerating organism death, with an enhanced effect in seawater, and that faecal streptococci are much better indicators than *Escherichia coli* of culturable pathogens, especially in seawater.

The near-shore levels of bacteria indicators, such as faecal coliforms, can be affected by distribution of discharged effluent under the influence of physical processes such as tidal state, wind velocity and density structure. Faecal coliform counts at certain sampling sites can differ by as much as two orders of magnitude according to the time of sampling (Milne 1989).

7.4.3 Chlorination

Chlorination if frequently used as a tertiary treatment in sewage treatment to kill surviving viable pathogens and to destroy residual organic material in the treated sewage effluent, including dead micro-organisms. Chlorination is an important method of disinfection in drinking-water treatment. Factors promoting survival of bacteria in chlorinated water supplies have been investigated by LeChevallier *et al.* (1988a,b). Several organisms grown in conditions of nutrient limitation, similar to those of the natural envionment and the final stages of low-rate sewage treatment processes, have been reported as being more resistant to disinfectants than those grown with high nutrient availability, including *Pseudomonas aeruginosa*, *Yersinia enterocolitica*, *Klebsiella pneumoniae* and *Legionella pneumophila.* Strains of bacteria producing an extracellular capsule have been isolated from chlorianted drinking-water, from which it has been surmised that the capsule protects the organisms from chlorine. It is also believed that attachment of bacteria to surfaces or entrapment in particles affords some protection from chlorine, as most viable bacteria in chlorine-treated drinking-water are attached to particles. The biofilm formed does not need to be thick and/or continuous, as sparsely distributed attached cells were found to be several hundred times more resistant to free chlorine than dispersed, suspended bacteria. Organisms have different susceptibilities to disinfection by free chlorine and monochloramine, so that disinfection acts as a selection mechanism, apparently favouring antibiotic-resistant strains.

The effects of selected disinfectants, sodium hypochlorite and monochloramine, were investigated with capsule-forming and non-capsule-forming strains of *Klebsiella pneumoniae* in attached growth and in suspended growth. Disinfection resistance was found to be increased by factors of 2 to 10 by encapsulation, biofilm age, temperature and medium used in previous growth conditions, and attached-growth culture of unencapsulated *Klebsiella pneumoniae* in high-nutrient medium increased disinfection resistant as much as 150 times. Biofilms grown for seven days were more chlorine resistant than those grown for two days. The effect of higher cell density in longer-established biofilms was discounted, as even the seven-day cell growth was sparsely distributed. The formation of an extracellular capsule did not in itself confer disinfection resistance, but did influence other resistance mechanisms. Unencapsulated *Klebsiella pneumoniae* grown in low-nutrient medium had double the chlorine resistance of those grown in high-nutrient medium, while capsule-forming *Klebsiella pneumoniae* grown in low nutrient medium had three times the chlorine resistance of those grown in high-nutrient medium. This was attributed to differences in the nature of the cell membrane and/or the capsule formed in the different growth conditions.

Disinfection by monochloramine was affected only by surfaces, while 'free chlorine' (hypochlorous acid) disinfection was affected by encapsulation, biofilm age and nutrients as well as by surfaces. An important discovery was that the resistance afforded by one mechanism could be multiplied by the resistance provided by an additional mechanism.

7.4.4 Irradiation

Investigation of the use of high-energy electron irradiation for disinfection of chlorinated and unchlorinated secondary wastewater showed that inactivation of indicator bacteria is a function of applied dose (Waite *et al.* 1989). The secondary effluent was the mixed outputs from a conventional and a dissolved oxygen activated sludge system, taken just before and just after chlorination prior to discharge.

With unchlorinated secondary effluent containing about 10^6 organisms/ml, at a flow-rate of about 7 l/s, a radiation dose of 600 krad gave a reduction of total coliforms of about 3 log-decades and total bacteria by 5 log-decades. Irradiation of the waste streams produced a substantial amount of hydrogen peroxide, which persisted in the wastewater for several days after irradiation. The amount of total oxidant formed increased linearly with radiation dose, giving 0.00045 N at 645 krad, equivalent to about 15 mg/l hydrogen peroxide.

With chlorinated secondary effluent, a dose of only 130 krad was required to give a reduction in total coliforms of 4 log-decades, but with no further decrease obtained with higher doses, even beyond 600 krad.

7.5 *CAMPYLOBACTER*

Campylobacter has only recently been recognized as an important cause of diarrhoea, and is now the single most frequently identified cause of gastro-intestinal infection in the UK, accounting for nearly 36 000 reported cases in 1989, or 34% of the total (DoE & DoH 1990). *Campylobacter* is a motile Gram-negative bacterium, with a polar flagellum, and is long and slender, less than 0.5 μm in diameter, in a curved or coiled form, and is sometimes described as a vibrio. Although *Campylobacter* cannot grow in atmospheric oxygen, it is not strictly anaerobic and grows at low levels of dissolved-oxygen tension (DOT) (Feachem *et al.*, 1983). *Campylobacter* are found worldwide in the intestinal and reproductive tracts of a range of wild and domestic animals and birds, including chickens, ducks, turkeys and seagulls. They are transmitted by contaminated water and food, particularly meat and unpasteurized milk, and infect humans by the faecal–oral route. *Campylobacter jejuni* is a major cause of acute gastroenteritis, while *Campylobacter foetus* can infect the blood, central nervous system and heart (Ketchum 1988). *Campylobacter* have considerable potential danger as a cause of diarrhoea. They are the most widespread of water-borne human pathogens, yet their occurrence is not dependent on the cycle between humans and domestic animals, as their natural hosts are wild birds and animals (Skirrow 1989). *Campylobacter* are likely to occur even in remote water sources and can survive for several weeks at temperatures below 15°C. Although they are readily eliminated by the standard water treatments used in the UK, outbreaks of *Campylobacter* enteritis resulting from faults in public water supplies have affected several thousand people, with attack rates as high as 50%.

Data on the survival of *Campylobacter* in faeces, urine, water and milk (Blaser *et al.* 1980 cited by Feachem *et al.* 1983) showed that a decrease of numbers of *Campylobacter* by a factor of ten million (seven log-decades) took three or four weeks at 4°C, three days at 25°C, and only 20 *minutes* in water acidified to pH 2.4 (at a temperature presumed to be 25°C). *Campylobacter* appears to be surprisingly sensitive to acid conditions, which suggests that the infective dose of *Campylobacter* must be very large for sufficient organisms to survive passage through the acid conditions in the stomach of the infected host.

7.5.1 Occurrence of *Campylobacter*

Bolton *et al.* (1987) found *Campylobacter* in samples from a river passing through urban and rural areas, in 43% of samples using membrane filtration and 21% using the 'most probable number' technique. Carter *et al.* (1987) found no correlation of the *Campylobacter* count with standard indicator bacteria in surface waters. *Campylobacter fetus* subsp. *venerealis* is an important venereal pathogen of cattle which can also be transmitted in artificial insemination in frozen semen (Chen *et al.* 1990).

Sporadic infection of *Campylobacter* enteritis at a university in the USA were investigated and three risk factors were identified: eating completely cooked chicken, eating raw or under-cooked chicken and contact with a cat or kitten. None of the subjects had drunk raw milk (Deming *et al.* 1987). *Campylobacter jejuni* was among the organisms found to be associated with episodes of diarrhoea in infants, together with enterotoxigenic and enteropathogenic *Escherichia coli*, *Shigella*, rotavirus and *Cryptosporidium* in an investigation of the incidence and aetiology of infantile diarrhoea and major routes of transmission in Huascar, Peru (Black *et al.* 1989). As previously mentioned in considering *Cryptosporidium* (section 6.2.2), an important route of infection was found to be in weaning foods, particularly as a result of improper preparation and cleaning of utensils, such as infection of sterile cans by a dirty can-opener, as well as poor domestic and personal hygiene generally.

In a two-year study of the distribution of 'thermophilic' campylobacters in human, environmental and food samples from the Reading (Berkshire, UK) area, with particular reference to toxin production and heat-stable serotype, 39% of food samples and 67% of environmental samples contained campylobacters (Fricker & Park 1989). Of the food samples, poultry (56%) and offal (47%) were commonly contaminated with *Campylobacter*, and sewage nearly always (97% of samples) contained campylobacters. There was a wide distribution of serotypes of *Campylobacter* in nearly all types of sample, and no correlation was found between serotype and toxin production. Not all strains associated with human enteritis produced toxin, as 23% produced cytotonic enterotoxin and 18% cytotoxin, while strains found in sewage frequently (18%) produced enterotoxin but seldom cytotoxin (5%), strains from river water seldom produced either type of toxin and a very few strains (less than 1%) produced both types of toxin. Production of enterotoxin and of cytotoxin were both more common in *Campylobacter jejuni* than in *C. coli.* It was concluded that there is a wide distribution of strains of *Campylopbacter* indistinguishable from those causing enteritis in humans, so that no one particular type of food can be designated as primarily responsible for human infections with *Campylobacter*.

7.5.2 *Campylobacter* in sewage and sewage treatment

A long-term study of occurrence, distribution and reduction of *Campylobacter* spp. in the sewage system and wastewater-treatment plant of a large town in Germany was reported by Höller (1988). The catchment area of the sewage-treatment plant included industrial, residential and argicultural areas and an abattoir, and the sewage was subjected to primary clarification and activated-sludge treatment before discharge to sea.

Over 4000 strains of several different species of *Campylobacter* were identified in the sewage, with *Campylobacter coli* predominating, especially in the abattoir effluent, and *Campylobacter jejuni* the next most common. This is unusual, as *C. jejuni* is usually excreted by humans and cattle. Other *Campylobacter* spp. found were *C. laridis*, *C. fetus* and a high proportion of non-typable *Campylobacter*. Most of the *Campylobacter* spp. in the raw sewage were identified as *C. coli* and *C. jejuni*, while most of those in the sewage sludge were classed as non-typable *Campylobacter*. Median counts of *Campylobacter* spp. were about 4000/100 ml in crude sewage, about 100/100 ml in aeration tank effluent and 20/100 ml in final effluent, showing that the sewage treatment process achieved 99.5% removal of *Campylobacter*. The *Campylobacter* counts in raw sludge varied widely, from virtually nothing to 10^6 cfu/100 ml, with a mean of about 100/100 ml. The mean value for activated sludge was about 1000 cfu/100 ml in a range 40 to 10^5 cfu/100 ml, and none were found in digested conditioned sludge. It was suggested that the low *Campylobacter* counts in the sludge samples could have been due to a transition to a non-culturable stage.

The occurrence of campylobacters in sewage and their removal by sewage treatment processes had already been investigated by Arimi, Fricker and Park in 1986. *Campylobacter* were enumerated by the most probable number technique in samples taken hourly over a five-day period from crude sewage, effluent from primary sedimentation and final effluent. The count of *Campylobacter* was reduced by 56% by primary sedimentation and by an overall 99.8% after secondary treatment using percolating 'filter' beds. Nearly all (97%) of the isolates indicated the presence of *Campylobacter jejuni*. Investigation of the inactivation of *Campylobacter jejuni* by chlorine and monochloramine showed that *Campylobacter jejuni* appears to be more susceptible to inactivation by chlorine and monochloramine than *Escherichia coli* (Blaser *et al.* 1986). By comparison with the response of *Escherichia coli*, *Campylobacter jejuni* and *Yersinia enterocolitica* were found to be more sensitive to ultraviolet light than many of the pathogens commonly associated with outbreaks of water borne disease (Butler *et al.* 1987).

7.5.3 *Campylobacter* and methaemoglobinaemia

Methaemoglobinaeemia is a rare disease in bottle-fed babies, resulting from the conversion of haemoglobin in the blood to methaemoglobin which is incapable of transporting oxygen. Oxyhaemoglobin, the oxygenated form of haemoglobin, is slowly converted to methaemoglobin so that about 0.5 to 2% of haemoglobin is normally in the form of methaemoglobin. If the proportion of methaemoglobin increases to about 10%, suffocation occurs and the baby's skin acquires a bluish tinge, known as cyanosis, so that the condition has become known as the blue-baby syndrome. Death may occur if the proportion of methaomoglobin exceeds 40%, but the condition responds rapidly to treatment, with a survival rate of about 70% and no reported after-effects (Bøckman & Bryson 1989). About 3000 cases of methaemoglobinaemia have been reported worldwide in the last 45 years, some of which were fatal. Of 1353 cases in Hungary between 1976 and 1982, 21 were fatal, but few cases have been reported in Western Europe or North America since 1970, and the last reported case in the UK was in 1978. In neonatal infants, 60 to 80% of their haemoglobin is of the foetal type, which is more susceptible to methaemoglobinaemia then adult haemoglobin. Haemoglobin can be converted to methaemoglobin by the action of certain chemicals, such as aniline, phenacetin and nitrite. There has been considerable concern about the risk of methaemoglobinaemia arising from contamination of water supplies by nitrate in agricultural run-off. In addition to being concentrated by heating during preparation of the baby's feed, it is believed that nitrate can be reduced microbiologically to nitrite in the mild pH conditions of the alimentary tract of a neonatal infant whos has not developed hydrochloric acid in its gastric juices.

It has been suggested that methaemoglobinaemia can result from enteritis alone, and has been implicated with infections by *Campylobacter jejuni* when the nitrate concentration in water was actually below 0.05 mg/l (Smith *et al.* 1988), and enterotoxic *Escherichia coli* with nitrate below 45 mg/l (Dagan *et al.* 1988). Some bacteria that cause enteritis, such as *Aerobacter aerogenes* and *Staphylococcus aureus*, can also reduce nitrate, so that water could need to be contaminated with sufficient nitrate and/or certain bacteria for methaemoglobinaemia to result. Most of the reported cases of methaemoglobinaemia involved well water that was contaminated microbiologically as well as with nitrate (Bøckman & Bryson 1989), although it can nevertheless hardly be said that the presence of nitrate is actually desirable.

7.6 *SALMONELLA*

Salmonella spp. are the bacteria that have been of most concern in the utilization of sewage sludge in agriculture, and were responsible for 26% of identified causes of gastro-intestinal infection in the UK in 1989, amounting to nearly 30 000 cases (DoE & DoH 1990). The nomenclature of *Salmonella* spp. is, unfortunately, somewhat confused, as a hangover from a system instituted 50 years ago, now discontinued, whereby each new isolate was named after the place where it was first identified. *Salmonella* spp. are Gram-negative, flagellate and motile rods and are facultative anaerobes. Infection occurs by ingestion of food or water contaminated by faeces from an infected creature, although there is evidence that infection can occur through breathing in aerosols containing the bacteria and even through contact with the eyes (Feachem *et al.* 1983). This, incidentally, is one reason why it is important for toilets to have lids, as flushing a toilet generates an aerosol, containing fresh faecal matter, which can persist for several minutes. In order to colonize the intestines, the bacteria have to pass through the acid environment of the stomach. For sufficient organisms to survive and cause infection, a very high infective dose of a million or so organisms may be necessary. The result of the infection depends on the type of *Salmonella* involved. The infection can be symptomless, while the host, known as a **carrier**, excretes millions of bacteria in each gram of faeces, unlike in other enteric bacterial infections. Certain serotypes, notably *S. typhi* and *S. paratyphi*, can invade the tissues, causing enteric fever, while other cause a typhoid-like disease or lesions on internal organs. (Feachem *et al.* 1983).

7.6.1 Typhoid fever

The most dangerous *Salmonella* infections involve *Salmonella typhi*, the organism causing typhoid fever. *Salmonella typhi* actually enter the bloodstream after colonising the intestinal tract and penetrating the intestinal mucosa, causing **bacteraemia** (bacteria in the blood) or, less specifically, **septicaemia**, continuous invasion of the bloodstream by micro-organisms. This results in acute fever, known as an *enteric fever*, and can be fatal. Diarrhoea is not a common symptom, although the faeces may contain as many as 10^{10} *Salmonella* organisms per gram. *Salmonella typhi* is a pathogen of humans and the only source of *Salmonella typhi* is a human carrying or infected with the organism. Typhoid fever is unusual in developed countries, but in Africa, Asia and Latin America, the WHO estimated there were a million infections resulting in 25 000 deaths a year in 1977 to 1978 (Warren 1988).

7.6.2 Other salmonellosis

The less dangerous, but nevertheless unpleasant and debilitating, *Salmonella* infection results in sickness and diarrhoea. Of 220 outbreaks of food poisoning reported in the USA in 1982, 55 were attributed to *Salmonella* spp., with milk, beef and poultry as the source (Ketchum 1988). These types of *Salmonella* are animal pathogens and are found in a wide range of animal species, from livestock and poultry to seagulls and insects. *Salmonella* organisms are especially prevalent in intensively reared poultry and livestock (Faechem *et al.* 1983) and raw meat the most likely carriers of *Salmonella* organisms causing food poisoning, as well as shellfish, milk products and salads (Ketchum 1988). Clearly animal slurries and slaughterhouse wastes constitute potential sources of *Salmonella* infection, and, as with *Cryptosporidium*, particular care must be taken to prevent contamination of water supplies with these materials as well as infection of people working with them. Infection of water reservoirs with *Salmonella* has been attributed to droppings from seagulls, which scavenge refuse tips and sewage-treatment works.

7.6.3 Persistence of *Salmonella* in sludge

As with other potentially infectious organisms, the survival of *Salmonella* in sludge or slurry spread on land depends on several different environmental factors. This leads to a correspondingly wide range in the bacterial survival rates reported. Assuming that bacterial die-off follows a first-order reaction pattern, the first-order death-rate coefficient, k_1, reported for *Salmonella* sp. was k_1=1.33 day^{-1}, or 5.5% dying **per hour**, in a range of k_1=0.21 to 6.93 day^{-1}, from published data. This compares with the mean half-life of over 120 bacterial species of 76 hours, corresponding to k_1=0.22 day^{-1} or 22% dying per day (Reddy *et al.* 1981 cited by Snowdon *et al.* 1989a). The death rate approximately doubles for every 10°C rise in temperature in a temperature range 5 to 30°C. The death rate of *Salmonella* in cattle slurry has also been reported as 90% dying in the first two to four weeks, with some surviving for as long as 150 days (Jones 1977 cited by Snowden *et al.* 1989a), which corresponds to k_1=0.08 to 0.16 day^{-1}, or 8% to 16% dying per day, with the original sample, of whatever size, containing 1.6×10^6 to 2.78×10^{10} viable cells. More than 99% of *Salmonella* can be removed by lagoon storage or cold anaerobic digestion (Pike 1986 cited by Snowden *et al.* 1989a). The effects of temperature and time on *Salmonella* have been summarized diagrammatically by Faechem *et al.* (1983) (Fig. 7.1).

7.6.4 *Salmonella* in mesophilic digestion

Mesophilic anaerobic digestion is the conventional method of sludge treatment, and its efficacy in destroying *Salmonella* was investigated by Pike *et al.* (1988). The inactivation of *Salmonella duesseldorf* during anaerobic digestion was assessed in laboratory-scale units fed daily, at 35°C, to give retention periods of 10, 16 and 20 days, and at 48°C with retention periods of 10 and 20 days. A decimal decay-rate coefficient k (days^{-1}) was determined from the relation

$$P/P_F = R/(10^{Rtk} - 1 + R) \qquad [7.1]$$

where R is the fraction of the working volume replaced at each feed, P and P_F are the organism counts in the treated and feed sludge respectively, and t is the mean retention period (in days).

With *Salmonella duesseldorf*, the decimal decay-rate coefficient was about 1 day^{-1} at 35°C and about 4 day^{-1} at 48°C, so that *Salmonella duesseldorf* added to raw sludge was inactivated in laboratory-scale anaerobic digestion three to four times more rapidly at 48°C than at 35°C, although the inactivation rate was to some extent influenced by the retention period. The concentration of *Salmonella* organisms in the feed was generally about 10^4 to 10^6/100 ml.

7.6.5 Irradiation of *Salmonella* in sludge

The effects of gamma radiation and subsequent storage on *Salmonella* and other bacteria in sewage sludge were reported by Carrington and Harman (1984b). Samples of primary sludge and surplus activated sludge from a plant treating

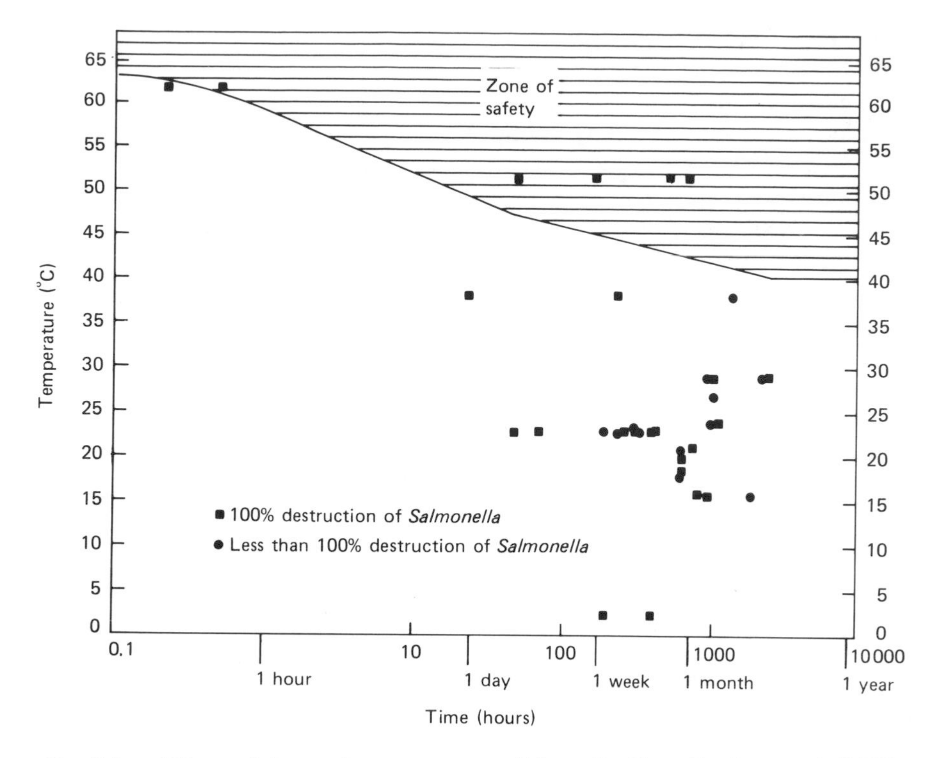

Fig. 7.1 — Effects of time and temperature on *Salmonella*. From Feachem *et al.* (1983). Reproduced by kind permission of The World Bank.

domestic sewage were inoculated with *Salmonella duesseldorf* and irradiated by exposure to cobalt 60 at doses between 0 and 600 krad. Bacterial counts in the control primary and secondary sludges respectively were coliforms, about 6 to 8 and 10 to 200×10^5/ml; viable bacteria about 4 to 300 and 30 to 4800×10^5/ml; spore-forming bacteria 2 to 350 and 10 to 55×10^4/ml; *Salmonella duesseldorf* 7 to 11 and 4 to 22/ml. The irradiation dose required to inactivate 90% of each population was 22 krad for coliforms, 76 krad for *Salmonella*, 188 krad for total viable bacteria and 192 krad for spore-forming bacteria. This showed that *Salmonella* were more resistant than coliform bacteria, but the aerobic, thermotolerant spore-forming bacteria showed the greatest resistance to irradiation. Coliform bacteria and *Salmonella* can be completely eliminated by radiation doses of only 200 krad, but to produce sterile sludge, a dose well above 600 krad is needed.

7.6.6 Health risks from *Salmonella* in sludge

Serotypes of *Salmonella* are very widespread in man and food animals, and are mostly zoonotic, transmitted naturally between animals and humans, rather than host-specific. The consumption of cooked meat, poultry and dairy products is the main means of transmission, and sludge is rarely involved. Cycles of infection in agriculture are sustained by intensive husbandry and the recycling of waste products.

The low infectivity of *Salmonella* for both man and cattle means that the establishment of clinical infection requires ingestion of several thousand viable cells. Salmonellosis is thus associated with heavy contamination, from the environment, or with contamination followed by multiplication, in contaminated food. There have been few cases of salmonellosis involving contamination with sewage sludge, and they could have been prevented by following the guidelines for the use of sewage sludge in agriculture (Bruce *et al.* 1990). A number of incidents of salmonellosis occurred before the publication of the guidelines, including ten incidents in the UK involving sewage effluents, eight with septic tank effluent, three with sewage sludge and two with abattoir effluent. One of these outbreaks involved the infection of 27 cattle and 98 people by raw milk, resulting from grazing cattle on pasture too soon after spreading chicken manure. The cause of the outbreak was identified through a rare strain of *Salmonella typhimurium* found in cattle, pasture, sludge and a chicken farm. Grazing animals on pasture too soon after spreading manure is a dangerous practice which appears to be on the increase. Leakage of septic tank waste from a household with a carrier of *Salmonella paratyphi* B contaminated a water supply and infected 98 people and five cattle, and raw sewage overflowing on to pasture from a cracked sewer pipe infected 90 cattle. In Switzerland, six outbreaks of salmonellosis involved the use of sludge on grassland (Havelaar & Block 1986, Reilly *et al.*, cited by Bruce *et al.* 1990).

Wastewater sludge samples were tested for the presence of *Salmonella* spp., *Shigella* spp. and *Campylobacter* spp. as part of a multi-year study of sludge application to farmland, the prevalence of bacterial enteric pathogens and the antibody status of farm families (Ottolenghi & Hamparian 1987). In a three-year period, 311 sludge samples from four wastewater treatment plants in Ohio, USA, were tested. Neither *Shigella* nor *Campylobacter* were detected, but *Salmonella* spp. were isolated 50 times, mostly from a treatment plant serving a large conurbation. The most frequently isolated serotype was *Salmonella infantis*. From the prevalance of antibodies to *Salmonella* in human subjects, it appeared that the risk of infection to an exposed population is minimal and no different from that of the non-exposed population.

7.7 *SHIGELLA*

Shigella spp. are Gram-negative, non-motile rods resembling *Escherichia coli* and *Salmonella*, and cause bacillary dysentery, mostly in tropical areas. *Shigella* are, however, prevalent throughout the world, and in 1989, over four thousand cases of shigellosis in the UK accounted for 4% of identified causes of gastrointestinal infection (DoE & DoH 1990). Of 220 outbreaks of food-poisoning reported in the USA in 1982, four were attributed to *Shigella* spp., with chicken or potato salad as the source of infection (Ketchum 1988). Transmission occurs by ingestion of food or water contaminated with faeces, and, although the infection can be symptomless, it can cause a severe, acute, debilitating diarrhoea or dysentery with cramps and vomiting. Flies are an important vector in the transmission of *Shigella*, leading to the mnemonic of the 'four Fs' for the recognized means of transmitting *Shigella*: food, faeces, fingers and flies. The bacterium colonizes the ileum and produces a protein exotoxin which inhibits protein synthesis in the host cells, like the diphtheria toxin.

As *Shigella* is Gram-negative, it has an endotoxin which can also cause diarrhoea and fever. The bacteria then invade the cells of the wall of the large intestine, producing inflammation and ulcers.

As with the other organisms discussed, the wide range of environmental factors affecting the survival of organisms in sludge or slurry leads to a correspondingly wide range in the bacterial survival rates reported. Assuming that bacterial die-off follows a first-order reaction pattern, the first-order death-rate coefficient, k_1, reported for *Shigella* sp. averaged 0.68 day^{-1}, corresponding to nearly 3% of the organisms dying per hour, in a range of k_1=0.62 to 0.74 day^{-1} (Snowdon *et al.* 1989a).

7.8 CHOLERA

Vibrio cholerae, the bacterium which causes cholera, is a Gram-negative facultative anaerobe which is transmitted by ingestion of large numbers of the vibrio in water or food contaminated with human faeces, although there are possibly reservoirs in surface waters free of faecal contamination. The cholera vibrio has a polar flagellum which enables the organism to move rapidly in liquid media. Cholera was the first disease scientifically linked to faecal contamination of water, in a classic example of detective work by John Snow during the outbreaks of cholera in London in the 1840s, although the organism responsible was not identified until forty years later.

Cholera results from a combination of invasive colonization of the intestines and the toxin produced by the bacteria. The whole of the intestinal tract, from jejunum to colon, is colonized by the cholera vibrios, which produce a heat-labile protein enterotoxin. The enterotoxin causes the intestinal mucosa to excrete chloride and bicarbonate ions, and the resultant electrolyte imbalance stimulates large quantities of sodium and potassium ions and water to enter the intestines. The resultant dehydration and loss of electrolytes eventually leads to death if untreated. The diarrhoea by the vast amounts of water entering the intestines is the product of the dehydration and contains large numbers of cholera vibrios. If these contaminate food or water supplies, the infection is transmitted further, and infection can be limited only by purification of water supplies and sewage. The simplest treatment for cholera is to restore the electrolyte balance *and* replace the lost water.

7.9 *ESCHERICHIA COLI*

Escherichia coli is a normal inhabitant of the intestines, but can be pathogenic outside and inside the intestinal tract. In 1989, 2% of identified causes of gastro-intestinal infection in the UK, amounting to over 200 cases, were attributed to *Escherichia coli* (DoE & DoH 1990). All strains of *Escherichia coli* can cause opportunistic infections of the urogenital regions, and certain strains carry one or more plasmids coded for one or more enterotoxins and for pili and cause dysentry. **Pili** are hair-like protein projections which enable the bacteria to attach to and colonize the intestinal mucosa.

Enterotoxic strains of *Escherichia coli* have pili *and* produce either a heat-labile enterotoxin similar to that in cholera, or a heat-stable toxin, or both. The bacteria cause disease when they both colonize the intestine and produce enterotoxin. The enterotoxin stimulates a flow of water and sodium and potassium ions into the intestines, resulting in dehydration, electrolyte imbalance and diarrhoea. This is sometimes known as traveller's diarrhoea, and the infection is acquired in areas with inadequate facilities for treating water and sewage.

Enteroinvasive *Escherichia coli* cause dysentry when the bacteria invade the cells lining the large intestine, in a similar manner to that of *Shigella*, causing ulceration and release of blood and pus into the faeces. This is much less common than enterotoxic infections. Diarrhoea can result from infection by a third strain of *Escherichia coli*, which is neither invasive nor toxin-producing, but whose mecahnism of causing disease is otherwise unknown.

7.10 *CLOSTRIDIUM PERFRINGENS*

The pathogenic effects of *Clostridium perfringens* as a food-poisoning organism have been succinctly reviewed by Granum (1990). *Clostridium perfringens*, a Gram-positive, non-motile spore-forming rod and an obligate anaerobe, is a normal inhabitant of soil and contributes to putrefaction processes. *C. perfringens* is found in dust, raw meat and the intestinal tract of man and animals, and is divided into five types classified according to their production of an enterotoxin and four major lethal toxins known as α-, β-, ε- and ι-toxins.

Type A: produces α- and enterotoxins, causes gas gangrene and food poisoning in man;
Type B: produces α-, β- and ε-toxins, causes dysentery in lambs and enterotoxaemia in sheep, goats and foals;
Type C: produces α-, β- and enterotoxins, causes necrotic enteritis in man and enterotoxaemia in calves, piglets, lambs and sheep;
Type D: produces α- and ε-toxins and may produce enterotoxins, causes enterotoxaemia in sheep, lambs and cattle;
Type E: produces α- and ι-toxins, causes enterotoxaemia in sheep and cattle.

To summarize the foregoing, all types of *C. perfringens* produce α-toxin; types A and C also produce enterotoxin and are pathogenic to man; types B, C, D and E are pathogenic to sheep, types D and E to cattle as well. Two types of *C. perfringens* which produce α-toxin and an enterotoxin are pathogenic to man; the others are all pathogenic to sheep, some to cattle or young animals.

C. perfringens has been reported in several countries as the most important cause of food poisoning, with only *Salmonella* infections having a greater incidence. Of 220 outbreaks of food poisoning reported in the USA in 1982, 10% were attributed to *C. perfringens*, all with meat of various types as the source (Ketchum 1988). Under adverse conditions, *C. perfringens* can form spores within the cell, called **endospores**, a highly resistant, dormant form which is very difficult to inactivate. Bacterial spores are the most heat-resistant forms of organism known, and can survive for years. Spores germinate into active vegetative cells in suitable conditions when stimulated by one of several treatments, which, rather ironically, are also used as disinfection methods in sludge treatment. Exposure to a temperature of 65°C, strong oxidizing agents or very acid conditions will stimulate spore germination, so that a material containing bacterial spores can have a higher bacterial count *after* disinfection than before. It should be noted that pasteurization destroys only *vegetative cells* of bacteria and not spores. This was the basis of a double pasteurization process called Tyndallization, whereby the first pasteurization inactivated vegetative bacteria and stimulated spore germination, and the second pasteurization inactivated the resultant vegetative bacteria.

C. perfringens causes two types of food poisoning, the mild type A food poisoning and the severe necrotic enteritis caused by type C, which is rare but often fatal.

C. perfringens Type A food poisoning is rarely fatal, and is caused by ingestion of food infected with 10^6 to 10^7 cells/g. The symptoms are due to the enterotoxin, which is believed to be part of the coat of the *C. perfringens* spore, although the strains associated with food poisoning produce as much as 2000 times more enterotoxin than that needed for spore-coat formation. Milligram quantities of the enterotoxin are required to cause food poisoning. Wild-type strains of *C. perfringens* isolated from natural sources do not produce sufficient enterotoxin to cause food poisoning. It is believed that excessive enterotoxin production results from disruption of control of enterotoxin synthesis caused by heating. This is an important point, as it suggests that incomplete heat treatment is worse than no heat treatment, as it actually *initiates* food poisoning. The enterotoxin is heat labile, and is denatured at 55°C, and is susceptible to acid and proteolytic enzymes. This indicates that food poisoning is caused by production of enterotoxin in the patient during sporulation of the organism, rather than by ingestion of pre-formed enterotoxin in food, which would be denatured on contact with the acid conditions in the stomach.

The rare but severe *Clostridium perfringens* type C food poisoning is due mainly to the β-toxin, although other toxins may have synergistic effects. The β-toxin is produced during vegetative growth and is degraded by proteolytic enzymes in the intestinal tract. Type C food poisoning then occurs by infection of nutritionally deprived patients, with a low protein intake that suppresses production of proteolytic enzymes or with a diet containing inhibitors of proteolytic enzymes, such as sweet potatoes (Granum 1990).

7.11 *CLOSTRIDIUM BOTULINUM*

Clostridium botulinum is a spore-forming, Gram-positive, rod-shaped obligate anaerobe found in soil and the intestines of animals. It produces a protein neurotoxin, whose action results in the disease known as botulism, which can be food-borne, a wound infection or undetermined (Telzak *et al.* 1990). The botulinum toxin is one of the most poisonous substances known, with the lethal dose for a human estimated as less than 1 μg. The protein toxin has a molecular weight of about 150 000, so that only one-thirtieth of a gram molecular weight would be enough to kill the entire population of the world. The toxin is heat-labile, but is acid-stable and resists degradation by digestive enzymes, unlike the toxins produced by *C. perfringens*. This means that food poisoning results from ingestion of food in which *C. botulinum* has been growing, and producing the toxin, rather than of the organism itself, unlike *C. perfringens*. The toxin is rapidly absorbed into the bloodstream from the intestines, and inhibits the transmission of nerve impulses to the muscles, causing paralysis. If the symptoms are recognized early enough, the toxin can be neutralized, otherwise death from suffocation results from paralysis of the respiratory muscles (Ketchum 1988).

Seven types of *C. botulinum* are identified on the basis of their antigenicially distinct toxins, of which four, types A, B, E and F are the principal causes of botulism in humans. Outbreaks of botulism in the UK are rare, but occur relatively frequently in the USA with about 50 cases a year. Of 220 outbreaks of food-poisoning reported in the USA in 1982, 21 were attributed to *C. botulinum* in meat, fish, fruit and vegetables (Ketchum 1988). This is because of the popularity of home-canning in the USA, where inadequate heat-treatment of contaminated food fails to inactivate clostridial spores, stimulates spore germination and provides a warm, anaerobic and highly nutrient environment ideal for the proliferation of *Clostridium* spp. bacteria. Outbreaks of botulism implicating fish products are associated with type E botulism, and nearly all cases of type E botulism have been traced to marine products. Salting, alone or in combination with sodium nitrite, sugar or smoking is commonly used to inhibit the growth of *C. botulinum*. In an outbreak of type E botulism, it was suggested that the viscera left in salted whitefish provided a low-salt environment that permitted growth and toxin production by *C. botulinum* (Telzak *et al.* 1990).

In the use of low pH levels to inhibit bacterial growth in foods, the inhibition effect depends on the nature of the acidulant used to produce the sub-optimal pH. In an investigation of the effect of citric acid on growth of proteolytic strains of *C. botulinum*, citric acid was found to be a million times more effective than hydrochloric acid in inhibiting growth of *C. botulinum* at pH 5.2 (Graham & Lund 1986). This is probably due to chelation of divalent metal ions, such as calcium or magnesium, by citrate, because the inhibitory effect of citric acid can be decreased by addition of calcium or magnesium ions.

7.12 *STAPHYLOCOCCUS AUREUS*

Staphylococcus aureus is a Gram-positive, facultative anaerobe found in the intestines and on human skin, and about 50% of strains produce an enterotoxin. Staphylococcal food poisoning is similar to that due to *Clostridium botulinum*, in that it is caused by ingestion of the toxin produced in food by the organism, rather than by the ingestion of the organism itself. Fortuntaly, the consequences are much less serious, and involve only sickness and diarrhoea, which pass in a couple of days. Food becomes contaminated when *Staphylococcus aureus* is transferred from the hands to the food during preparation, and the bacteria proliferate if the contaminated food is then kept warm. The enterotoxins produced by *Staphylococcus aureus* are heat-stable and so may not be destroyed by subsequent re-cooking of the contaminated food, and are resistant to acid and enzymes in the stomach. There were 28 outbreaks of staphylococcal food poisoning reported in the USA in 1982. Many different toxic substances are excreted by various strains of staphylococci, and certain strains of *Staphylococcus aureus* can produce an exotoxin causing Toxic Shock Syndrome, which can be fatal (Ketchum 1988).

7.13 YERSINIOSIS

Yersinia enterocolitica is a Gram-negative rod roughly 2 μm long and 1 μm in diameter, which can appear ovoid. The genus *Yersinia* includes *Y. pestis*, the organism causing bubonic plague, spread by fleas from rats dying of plague. *Y. enterocolitica*, however, has been recognized as a human pathogen only recently. *Y. enterocolitica* causes gastroenteritis in humans, and appears to be similar to salmonellosis in its reservoir and transmission, but it can cause septicaemia and the resultant fever (Feachem *et al.* 1983). It is likely that sludge treatment processes that inactivate *Salmonella* satisfactorily will also inactivate *Yersinia* spp.

7.14 LISTERIOSIS

The food-poisoning bacterium *Listeria monocytogenes* can cause septicaemia and meningoencephalitis in humans and mastitis in cows. Pregnant women are particularly at risk, because *Listeria monocytogenes* can cross the placenta and infect the foetus. This can result in premature birth, deformity or even still-birth or abortion of the foetus. *Listeria monocytogenes* is found in sewage, soil, water and animals, but has most recently been associated with dairy products, particularly soft cheeses. Infection occurs as a result of ingestion of milk from an infected cow, or cheese produced from infected milk, or direct contact with an infected animal.

7.14.1 Transmission of listeriosis

A factor recognized as important in the transmission and epidemiology of *Listeria* infection is prolonged excretion of *Listeria monocytogenes* in milk. In an investigation of this factor, populations of *Listeria* spp. in sources of raw milk were found to be not particularly persistent (Slade *et al.* 1989). *Listeria* organisms of various species were found in over 36 out of 315 samples of raw milk sampled from bulk tanks in Canada. Of the infected sources, 34 were available for re-sampling, from 21 of which *Listeria* were re-isolated a month or more later, and in only one more than five months later. It was presumed that sources were periodically recontaminated from the cattle habitat, and absence of manure has been determined as an important control factor, with fluctuations of strains and species of *Listeria* attributed to feral animals and birds.

7.14.2 *Listeria* in sewage and sludge treatment

Listeria monocytogenes occurs and survives for long periods in sewage, survives sewage treatment and occurs in treated effluent and in dried sludge cake used as fertlizer, survives longer than *Salmonella* on land sprayed with sludge, and has been isolated from vegetation and contaminated crops. In an investigation of the effects of sewage treatment on the removal of *Listeria monocytogenes* from sewage, *Listeria monocytogenes* was found in samples taken from all stages of sewage treatment in two sewage-treatment plants near Baghdad, Iraq, including the final effluent and the dried sewage sludge cake (al-Ghazali & al-Azawi 1988a). The principal reduction in numbers (85 to 99.7%) occurred after the aeration and anaerobic digestion (30 days, presumably at ambient temperature) stages. Counts were generaly less than 20 to 30 cfu/ml, but were occasionally over 1000/ml. From an investigation of the effects of storage of sewage sludge cake on the survival of *Listeria monocytogenes*, the same authors reported that the counts of *Listeria monocytogenes* in a 5-ton heap of sewage sludge cake left to dry in the sun in Iraq were reduced from less than 10/ml to undetectable levels after eight weeks, and had not reappeared after a further two months (al-Ghazali & al-Azawi 1988b). The effect was attributed to the reduction of the moisture content from an initial value of about 40% to about 10% within the heap after eight weeks. For the first eight weeks, the pH was neutral or slightly alkaline (7 to 7.5).

7.15 *AEROMONAS*

Aeromonas spp. have recently become recognized as human pathogens responsible for gastroenteritis, cellulitis and bacteraemia, and, although they are not considered to be of faecal origin, the number of aeromonads in natural waters are increased by sewage pollution (Araujo *et al.* 1989). Aeromonads in drinking water have been associated with gastroenteritis in children, and *Aeromonas* spp. have been found to produce cytotoxins (Versteegh *et al.* 1989). The possible relation between *Aeromonas* and faecal coliforms in fresh waters was investigated in rivers in the Barcelona area of Spain (Araujo *et al.* 1989). The results showed that there is indeed a significant correlation between the counts of aeromonads and faecal coliforms and the concentration of organic nutrients, expressed as biologically-determined oxygen-demand (BOD), in polluted waters. There was, however, no correlation between numbers of *Aeromonas* spp. and faecal coliforms in waters free from faecal pollution. In other words, there is a close relationship between *Aeromonas* spp. and faecal contamination indicator organisms where organic nutrients are of faecal origin, but not where organic nutrients are of non-faecal origin. Thus, using faecal indicator bacteria as the only measure of infection risk underestimates the risk due to opportunistic pathogens such as *Aeromonas* spp. It is therefore advisable to monitor *Aeromonas* spp. in water used for recreation, drinking and aquaculture. Aeromonads have been found sensitive to copper concentrations as low as 10 μg/l, picked up from a copper sampling pipe, although coliform bacteria were less sensitive. The copper sensitivity was neutralized by addition of 50 mg/l disodium EDTA, although the effect on faecal streptococci was less, particularly at high pH levels (Versteegh *et al.* 1989).

8

Viruses in sludge

8.1 INTRODUCTION

A wide range of viruses exist in sewage, and well over a hundred different types of virus of human origin have been identified in wastewaters, and some may be present in considerable numbers. By far the largest group of viruses in wastewaters are of human origin (Grabow 1968), but viruses of higher plants and animals, and, of course, bacteriophages are also found in sewage. Faeces are not the only source of human viruses, as nose and throat secretions, containing respiratory viruses, and washing- or bath-water, containing viruses from skin lesions, and blood, also find their way into sewage. Faeces are estimated to contain 10^6 to 10^{10} virus particles per gram (Yates & Yates 1988). In raw sewage, 100 plaque-forming units (pfu) per millilitre have been observed, septic-tank effluent has yielded 800 to 7500 pfu/ml, and in groundwater levels of 0.004 to 0.025 pfu/ml have been reported. The infective dose is believed to be low and has been estimated as one to 200 virus particles (Snowdon *et al.* 1989a).

Viruses affecting plants and animals are of economic importance, since disposal of wastes containing viruses to agricultural land would involve a risk of disseminating disease. Wastes from farms, slaughterhouses and food-processing factories are likely to contain the highest numbers of plant and animal viruses. Once applied to land, however, the virus count is reduced by exposure to adverse environmental conditions, particularly temperature and desiccation to which viruses appear to be especially sensitive. The nature of the sludge itself and the treatment to which it has been subjected are decisive factors, and are discussed in more detail further on.

The transport of viruses from sludge application sites was reviewed by Jørgensen and Lund in 1985 (Jørgensen & Lund 1986). Sludge, even after digestion, may contain 10 000 to 100 000 infectious virus particles per litre. Viruses in sludge applied to land may be inactivated at the site of application, or may be transported by water-flow horizontally to other areas of land or into waterways, or vertically into the soil and groundwater. Although human enteric viruses have often been found in surface waters, this is generally attributed to direct contamination by sewage discharges, rather than from land application of sludge. Vertical transport of viruses through soil depends on the permeability of the soil as well as the other environmental conditions. Viruses may be inactivated during the transport process, for example by passive adsorption and desorption, but can be very slow. Viruses may be carried in aerosols created when sludge is sprayed on to land, although investigations in the USA found no significant correlation between illness and exposure to sludge aerosols. This is probably because viruses are associated with the solid content of sludge, and aerosol particles large enough to contain solid particles are likely to be too heavy to be carried very far.

8.2 PROPERTIES OF VIRUSES

8.2.1 Assessment of viruses

Viruses are grown in cultures of susceptible cells, and the presence and growth of viruses are observed as a **cytopathic effect** (cpe) resulting from alteration or destruction of the cells. Virus infectivity can be assayed using replicate monolayers of susceptible cells, inoculated with samples of serial dilutions of the virus-containing medium being tested. The end-point of the test is then the highest dilution whose samples cause a cytopathic effect. The cytopathic effect test is useful for detecting the presence of infective viruses, but enumerating infective virus employs a modification of this test. The spread of newly synthesized virus particles is restricted by covering the monolayer of susceptible cells with a layer of nutrient agar. Where the growth of virus causes cell lysis, a circular hole, called a **plaque**, is then formed in the cell layer, which is revealed by staining the residual living cells. The number of infective virus in the original medium is calculated from the number of plaques formed, the dilution and the inoculum size. Numbers of virus are thus often expressed as **plaque-forming units** or **pfu**'s, analogous to colony-forming units in counting bacteria. Plaque assay has the disadvantage of taking several days to produce a result. Rapid methods utilizing the antigenic properties of viruses have now been developed, which give an indirect measure of the *number* of virus, but no indication of *infectivity*.

Not all viruses can be cultured *in vitro*. Those that can be cultured in the laboratory are known as **culturable** viruses, and those that cannot are called non-culturable or **fastidious** viruses.

8.2.2 Properties of viruses

Viruses are obligate intracellular parasites, with a very limited range of hosts. As viruses are replicated only by their host cells, they will never increase in numbers once they are released into the environment away from their host, but will generally decrease or possibly persist. Viruses disseminated by the faecal route are termed **enteric viruses**, and care should be taken to avoid confusion with the term **enteroviruses**, which is the name of one important group of enteric viruses. As their host environment is the gastro-intestinal tract, a characteristic of enteric viruses is their ability to withstand acid conditions with a pH level of about 3. A two-year study of enteric viruses in wastewater at a sewage-treatment plant in Israel (Buras 1976) showed that enteric viruses were present in wastewater throughout the year, but fluctuated with the greatest numbers occurring in summer. The prevalent viruses in the winter period were poliovirus 1 and echovirus 4.

8.2.3 Nomenclature of viruses

The nomenclature of viruses is currently somewhat confusing. The International Committee on Nomenclature of Viruses has developed a system based on the structure, size and chemical composition of viruses. The old vernacular classification according to the host the virus infects and the resulting disease, or the place where the disease was first identified, is still, however, widely used. In this review, to avoid error, the nomenclature used will be that employed in the original publication, and nearly all those cited here use the vernacular nomenclature. An informal convention of terminology is that the plural of 'virus' is 'virus' when the plural signifies a number of one specific type of virus particle, but is 'viruses' when the plural signifies a number of different types of virus.

8.2.4 Structure of viruses

The essential component of a virus is a quantity of nucleic acid coded with the instructions for synthesizing the virus, which can be either DNA or RNA. There are six families of DNA viruses and 11 of RNA viruses. When a virus particle penetrates a suitable host cell, the viral nucleic acid takes over control of the cell metabolism so that the cell produces more virus. The viral nucleic acid can be in several different forms in different viruses. The DNA or RNA can be single- or double-stranded, as a single molecule or in several segments, as an endless loop or as a straight chain with two uncombined ends. The nucleic acid in most viruses is in the form of a single molecule of either DNA or RNA, either single- or double-stranded (Ketchum 1988). A detailed account of the structure of viruses is beyond the scope of this discussion, but some basic features play a significant part in the removal of viruses from wastewaters or inactivating them in treatment processes.

The nucleic acid is surrounded by a proteinaceous shell called a **capsid**. A simple virus can consist only of a proteinaceous capsid containing nucleic acid. Some simple viruses have a double protein shell. In more complex viruses, the capsid may be contained within a membrane, called the **viral envelope**, and enzymes can be contained in the envelope with the nucleocapsid. A virus without an envelope is known as a **naked** virus. The viral envelope is derived from a membrane in the infected host cell, such as the endoplasmic reticulum. This is important, because cell membranes contain lipids as well as proteins, and can thus be disrupted by detergents, such as sodium dodecyl sulphate, and non-polar solvents such as ether, although enteric viruses are resistant to ether. Detergents are, of course, very likely to be present in wastewater. The effects of detergents on naked viruses is not easy to predict, as, for example, detergents have been shown to increase the rate of heat inactivation of reoviruses, while a heat-stable ionic detergent protects poliovirus against thermal inactivation (Bitton 1980).

The viral nucleic acid core is affected by formaldehyde, nitrous acid and ammonia (Bitton 1980). Formaldehyde is widely used as a disinfectant, ammonia is normally present in sewage and nitrous acid is formed in aerobic wastewater treatment by **nitrification**, the microbiological oxidation of ammonia. Ammonia also diminishes the protective effect of detergents on poliovirus. The toxic effects of ammonia in microbiological systems are due to free, unionized ammonia, the proportion of which increases with increasing pH and temperature. It is thus likely that this contributes to the virucidal effect of high pH levels, and ammonia has been proposed as a preferable alternative to chlorine for disinfecting sludge (Winkler 1984).

Animal viruses range in size from 20 to 30 nm to over 200 nm. Virus particles are thus too small to separate by sedimentation in wastewater treatment processes, unless they can be made to form very large aggregates or attach to larger particles. Like other protein materials, the proteinaceous viral capsids can be utilized as nutrients by bacteria and other organisms in wastewater, after proteolysis catalysed by microbial enzymes. With such a wide range of differences among different types of virus, it is likely that a treatment method found to work successfully with one type of virus will not necessarily be as effective with other types.

8.2.5 Bacteriophages

Bacteriophages are viruses that infect bacteria. Much of the fundamental research on viruses involved bacteriophages, because it is much easier to grow bacterial cultures, in which they reproduce, than animal cell cultures. Interestingly, one of the first bacteriophages discovered was one that destroys the dysentery bacillus, *Shigella*. Bacteriophages are different from animal viruses, however, because bacteria are enclosed by cell walls as well as cell membranes, whereas animal cells have only a cell membrane. Bacteriophages thus need to be equipped with the means of penetrating the cell wall as well as the cell membrane. Some bacteriophages have a **tail** which acts as a microscopic (or 'picoscopic', perhaps) hypodermic needle to inject the nucleic acid into the bacterial cell. The destruction of bacteria by bacteriophage infections in industrial fermentations can result in serious financial losses: a single lost industrial-scale fermentation batch can cost several hundred thousand pounds.

8.2.6 Viroids and prions

Nucleic acid alone can be infectious, in the form of small single-stranded loops of RNA called **viroids**, which infect and multiply in plant tissues. Potato spindle tuber disease is caused by a viroid, for example. Infectious proteinaceous agents called **prions** are suspected of being the cause of slow-developing neurological diseases of animals, such as scrapie in sheep, bovine spongiform encephalopathy ('mad cow disease') and Kreuzfeld–Jakob disease in humans. If this suspicion is in fact correct, then prions in animal slurry or sludge derived from sewage containing slaughter-house wastewater constitute a very serious problem in sludge treatment. This is because prions are even more resistant to adverse conditions than bacterial endospores (section 7.1.3). Heating at 121°C for over four hours or repeated heating in molar caustic soda is required to inactivate prions (Ketchum 1988).

8.3 VIRAL INDICATORS

As with other parasites, the presence and numbers of viruses in environmental systems do not correlate well with the standard bacterial indicators of faecal contamination, particularly in water with a very low level of contamination. Bacteria may be able to reproduce outside the host, whereas viruses cannot, and vegetative bacteria are generally less resistant to disinfection than viruses.

A distinction also needs to be made between indicators and tracers. An indicator organism needs to be characteristic of faecal contamination as well as readily assayed. A tracer organism is one which is used to test the effects of treatments on organisms. Tracers need to be easily assayed and have survival properties similar to those of the organisms of interest. Ideally, the best possible tracer would also be a reliable indicator. Enteric viruses pose a particular problem, however. It happens that viruses that cause gastroenteritis are difficult or impossible to culture *in vitro* (Adams & Lloyd 1989), so there is thus much effort expended in the search for an indicator virus that is safe to handle, straightforward to detect and count, and whose behaviour models that of important pathogenic viruses. The immunoperoxidase procedure now enables assays of non-culturable rotavirus to be determined in less than 24 hours. As rotavirus is an enteric virus resistant to adverse conditions, it could act as a valuable rapid viral indicator.

Bacteriophages are regularly investigated for use as indicators, rather than for their effects on their host bacteria in sewage. It should be borne in mind, however, that viruses have a wide range of properties, and have a correspondingly wide range of resistance to treatment processes. The size of the virus would affect its susceptibility to irradiation, for example (sections 2.2.7 and 8.9.2), and a virus with an envelope would be expected to be more resistant to irradiation than a naked virus.

Indicator systems for assessment of the virological safety of treated drinking water have been reviewed by Grabow (1986), who concluded that waterborne outbreaks of viral disease were virtually all caused by water that did not meet conventional *bacteriological* standards. The technology for the treatment and quality surveillance required for the production of virologically safe water is available, but that it often applied inefficiently.

It has been suggested (Havelaar & van Olphen 1989) that culturable enteroviruses and reoviruses should be used as indicator organisms. Although they are not associated with water-borne viral disease, they are much more easily cultured than, for example, viruses causing gastroenteritis and infectious hepatitis. In addition, entero- and reoviruses are persistent in water and treatment processes, so that disinfection procedures removing these will also deal effectively with epidemiologically important viruses.

A type of virus proposed as an indicator of viral contamination in water is the F-specific RNA-bacteriophage (Havelaar & Pot-Hogeboom 1988). F-specific RNA bacteriophages, 'fRNA-phages', belong to the Leviviridae family, and comprise a single strand of RNA in a protein capsid, with a diameter of 20 to 27 nm. They are thus similar in size and structure to the enteroviruses, and this physical resemblance suggests the use of fRNA-phages as enterovirus models. A model organism should have an ecology and resistance to inactivation similar to that of the pathogen it models, but to be more easily and/or reliably assayable. The f2 and MS2 strains have been shown to be resistant to inactivation in the natural environment, in sewage-treatment processes, by heat, disinfectants and radiation, although they appear to be susceptible to high-powered oxidizing agents such as hypochlorous acid, ozone and peroxy-compounds.

The fRNA-phages infect their host cells after preliminary attachment to the sides of the F- or sex-pili produced by *Escherichia coli* or related bacteria possessing the F-incompatibility plasmid. The F-pili are produced only at temperatures above 30°C, or if the host cell has been previously grown at 37°C, as in the intestines of a warm-blooded animal. The fRNA-phages are thus unlikely to multiply in the ordinary environment, and the presence of fRNA-phages is thus indirectly associated with faecal pollution.

Investigation showed that fRNA-phages were absent from the faeces of humans, dogs, cows and horses, and were present in only low numbers in the faeces of pigs and calves. High counts of 10^3 to 10^7 pfu/g were found in the faeces of 'broiler' chickens and 10^3 to 10^4 pfu/ml in several different types of wastewater, including domestic, hospital, slaughterhouse and poultry-processing wastewaters.

Measurements of the effects of chlorination at about 1.5 mg/l on *Escherichia coli*, fRNA-coliphage and enteroviruses in secondary sewage effluents showed that the enterovirus counts were followed much more closely by those of the fRNA-coliphage than by those of the *E. coli* (Havelaar & van Olphen 1989). In addition, the fRNA-coliphage appears to be more resistant to chlorination at that level than the enteroviruses. Tests determining the dose of ultra-violet radiation giving 99.9% inactivation indicated that fRNA-phages f2 and MS2 are somewhat more resistant to ultra-violet than poliovirus 1, coxsackievirus B1 and reovirus, and considerably more resistant than *E. coli*. It is also suggested as a more convenient, as well as a more realistic, organism for calibrating ultra-violet treatment systems than a suspension of *Bacillus subtilis* spores.

The two diseases principally associated with viruses in sludge are gastroenteritis and viral hepatitis.

8.4 VIRAL GASTROENTERITIS

Enteric virus infections are most frequent in infants and young children, and transmission of enteric viruses to other children and adults occurs directly or indirectly via the faecal–oral route. Land application of human waste may result in indirect transmission of enteric viruses by flies, food or water, by contamination of food crops, contamination of the water supply by surface run-off or through groundwater or by infection of workers dealing with the wastes (Snowdon *et al.* 1989a). Viruses enter the water cycle through faecal contamination, and without properly controlled water-treatment processes, and disinfection in particular, they may enter the drinking water supply (Morris 1989). The negligent disposal of sewage may therefore lead to the dissemination of viral disease. An IAWPRC study group concluded in 1983 that not only was there no doubt that pathogenic viruses were transmitted by water, but that available data on water-borne viral disease underestimated the true extent of the hazard. The report also drew attention to the fact that the technology to produce virologically safe water existed but was insufficiently applied.

The enteric viruses comprise over 100 different viruses known to be excreted in human faeces and are routinely isolated from sewage-polluted water. The principal types include the laboratory-culturable polioviruses, coxsackie viruses, echovirus, adenovirus and reovirus, which replicate in the cells of the intestine and can be excreted in numbers over a million per gram of stool. Although these culturable viruses are not usually associated with gastro intestinal symptoms, they can give rise to other symptoms ranging from skin rashes to conjunctivitis and meningitis. The 'fastidious' non-culturable viruses, such as Norwalk and rotavirus, are associated with acute diarrhoeal disorders. Although these are also shed in very large numbers, about 10^{10}/g (faeces), these viruses can be detected only by using sophisticated methods such as electron microscopy or immunofluorescent antibody techniques (Merrett *et al.* 1989).

Viruses are actively shed in the faeces by a person undergoing an enteric virus infection, during the course of the infection. This is not necessarily accompanied by gastro-intestinal discomfort, and at any time, a few per cent of a population may be experiencing enteric virus infection.

It would be expected that the most important groups of viruses found in sewage would be those associated with gastro-intestinal disorders. Their accumulation in sludge and their survival in sludge treatment are therefore a vital consideration in the use of sewage sludge. Several groups of viruses are associated with gastroenteritis, including caliciviridae, the picornaviridae poliovirus, echovirus and coxsackieviruses A and B, the parvoviridae adenovirus and parvovirus and reoviridae reovirus and rotavirus.

The level of virus contamination causing disease has proved to be very difficult to assess. Although one or two enteric virus plaque-forming units may be sufficient to cause infection, infection does not necessarily result in disease. Experiments with healthy adults indicated a minimum infective dose of 17 pfu for the very mildly pathogenic echovirus 12 in drinking water and 1 pfu of rotavirus. Infected individuals can then pass the infection on to subjects who have had no contact with the original contaminated medium, a well-documented effect with, for example, Norwalk virus. Some of the enteroviruses are well-known for giving rise to symptomless infections, and the development of clinical illness depends on a range of factors including the immune status and age of the host, the type, strain and virulence of the infecting organism and the infection route. For individuals infected with hepatitis A virus, only a few per cent of children show clinical illness, and the proportion increases with age, while the opposite trend is observed with rotavirus infections. The mortality risk for hepatitis A virus is 0.6% in the USA and for other enterovirus infections ranges from less than 0.1% to about 1.8% (Gerba & Goyal 1988).

In a survey of epidemiological studies of people in contact with wastewater and sludge (Block & Schwartzbrod 1982 cited by Schwartzbrod *et al.* 1987), infections with viruses including hepatitis A virus and coxsackievirus B3 from sewage, rotavirus and Norwalk virus among sewage workers were reported. A number of unspecified virus infections were also reported among sewer workers and farm workers.

8.5 ENTERIC VIRUSES

A brief account of salient features of important enteric viruses is given in this section.

8.5.1 Reoviridae

8.5.1.1 Rotavirus

In the UK in 1989, 14% of identified causes of gastro intestinal infection were attributed to rotavirus, amounting to nearly 15 000 cases (DoE & DoH 1990). **Rotavirus**, classified in Reoviridae, is a naked RNA virus with a wheel-shaped double-shelled capsid about 70 nm in diameter. Rotaviruses are associated principally with acute infant diarrhoea, whose symptoms can range from mild diarrhoea to severe gastroenteritis that can actually be fatal. The severe form is most likely to affect infants between six months and two years old, and death can result from dehydration and the resulting electrolyte imbalance. The number of virus in faeces can rise to as much as 10^{10} rotavirus particles per gram for a few days. Infection occurs via the faecal–oral route, occurring throughout the year in tropical climates and in winter in temperate climates. Ketchum (1988) pointed out that rotavirus infection is so widespread that preventing the spread of the disease cannot be achieved by adequate sewage treatment alone. Rotaviruses are carried by the young of a wide variety of domestic animals, including mice, rabbits, lambs and calves. A very high level of rotaviruses has been found in sewage in Barcelona, Spain, up to 14 000 fluorescent foci per litre, and similar to that of enteroviruses, but no correlation with the level of human enteric viruses was found (Bosch *et al.* 1988).

An assay procedure recently developed for rotaviruses may enable them to be used as indicators of the presence of enteric viruses (Adams & Lloyd 1989). Of the 100 or more viruses excreted by humans, those that cause gastroenteritis are difficult or impossible to culture. Readily cultured viral indicators are therefore needed to mimic the occurrence of non-culturable viruses in polluted waters and shellfish. Conventional extraction and enumeration procedures even for culturable viruses are time-consuming and their results take several days to obtain, but the immunoperoxidase procedure now enables assays of rotavirus to be determined in less than 24 hours. As rotavirus is an enteric virus resistant to adverse conditions, it could act as a valuable rapid viral indicator.

8.5.1.2 Reovirus

Reovirus are found in polluted water, but it is not clearly established what disease or symptoms they cause. They are naked viruses, with double-stranded RNA in a double protein shell, about 80 nm in diameter.

8.5.2 Caliciviridae

Norwalk virus, classified in Caliciviridae, is a naked RNA virus, about 30 nm in diameter, with a nucleocapsid. Norwalk agents form a group of viruses causing epidemic diarrhoea and acute non-bacterial gastroenteritis, first observed in gastroenteritis patients in a town called Norwalk in the USA. These viruses are presumed to be transmitted by the faecal–oral route, but they are virtually impossible to culture *in vitro*. It has been estimated that viruses may be responsible for half the outbreaks of water-borne diseases in the USA, of which Norwalk and Norwalk-like viruses may be responsible for about half (Yates & Yates 1988). Several outbreaks of water-borne disease in the USA were caused by contamination of groundwater with viruses present in septic tank effluent. Other viruses found to cause similar diarrhoea in humans and found in the host's faeces include coronaviruses, other caliciviruses and astrovirus (Rao & Melnick 1986).

8.5.3 Picornaviridae

The enteroviruses are classified in the picornaviridae, from *pico* (small) and RNA, and are about 30 nm in diameter with a naked, icosohedral capsid. This group includes the polioviruses, echovirus and coxsackie A and B viruses. The hepatitis A virus (HAV) is also a picornavirus, but, in order to contrast its effects with other hepatitis viruses which are not picornaviruses, it is considered separately in section 8.6. The pathogenic effects of enteroviruses have been summarized (Rao & Melnick 1986) as causing very serious diseases in only a small proportion of infections.

8.5.3.1 Coxsackieviruses A and B

Coxsackievirus and **echovirus** are naked RNA viruses with a 30-nm capsid. Another town in the USA has given its name to a group of viruses known as the coxsackieviruses, divided into two groups, A and B. A similar group of viruses giving rise to a similar range of symptoms as coxsackieviruses are enteric cytopathic human orphan viruses, known by the acronym echovirus. They were found to cause cytopathic effects, changes in cells in tissue culture, and are known as 'orphans' because they can infect humans without giving rise to clinical symptoms. Infection occurs by the faecal–oral route, and although infections by all types of coxsackievirus and echovirus can be symptomless or cause only a mild fever, some types can cause encephalitis and aseptic meningitis. Certain types of coxsackievirus A and B and echovirus are responsible for about half the cases of aseptic meningitis, which can also be caused by poliovirus. Certain types of coxsackievirus A and B can cause a disease producing lesions on the hands, feet and mouth.

8.5.3.2 Poliovirus

Poliovirus is also classified in the picornaviridae with coxsackievirus and echovirus, and is a naked RNA virus about 20 to 30 nm in size. Polioviruses cause water-borne viral disease and are transmitted by the oral–faecal route and by respiratory secretions. The infective dose is possibly as small as a single virus particle. Nearly all infections are symptomless, although infected individuals excrete large numbers of poliovirus in faeces, about a million per gram, with a concomitant risk of the infection spreading. In less than 10% of cases, symptoms of fever, vomiting or diarrhoea may occur for a few days, and in about 1 or 2% of cases, the virus penetrates the nervous system and causes meningitis and paralysis, and possibly even death. Coxsackieviruses and echoviruses can also occasionally cause poliomyelitis (Feachem *et al.* 1983). There have been surprisingly few outbreaks of poliomyelitis attributed to contaminated water, considering that poliovirous are detected in sewage in large numbers (Rao & Melnick 1986). The WHO estimated that in Africa, Asia and Latin America, there were 80 million poliomyelitis infections and ten to twenty thousand deaths per year in 1977 to 1978 (Warren 1988).

8.5.4 Parvoviridae

Adenoviruses are found in faeces and sewage and can cause respiratory and eye disease as well as gastroenteritis. Enteric adenoviruses have been found in faeces of up to 12% of young children with acute gastroenteritis (Rao & Melnick 1986). Adenoviruses are naked and about 80 nm in diameter, with double-stranded DNA in a proteinaceous capsid. Very small viruses about 20 nm in diameter, associated with adenoviruses, have been recovered from faeces, and are known as **parvoviruses** or adeno-associated viruses (AAV). They are thought to be associated with respiratory disease in children (Rao & Melnick 1986, Feachem *et al.* 1983).

8.6 VIRAL HEPATITIS

The literal meaning of *hepatitis* is 'inflammation of the liver', and can range in severity from mild to fatal. Hepatitis generally results in jaundice and/or dark urine, sickness and fever. Several different unrelated viruses cause hepatitis, some of which are still not completely identified. As a result, the nomenclature used for viruses causing hepatitis is unsystematic, with the hepatitis viruses designated by letters of the alphabet.

Hepatitis A virus (HAV) is the cause of infectious hepatitis. HAV is the hepatitis virus of most concern in sludge handling and treatment, because it is shed in the faeces of infected hosts over an extended period, and is transmitted by the faecal–oral route, mainly person-to-person contact. HAV can also be transmitted by contaminated water and shellfish grown in contaminated water. HAV has proved difficult to inactivate, requiring heat treatment of 30 minutes at 70°C or treatment with formaldehyde at a concentration of at least 0.35% (Flehmig *et al.* 1985). HAV is a naked, spherical RNA virus about 25 to 30 nm in diameter, and is thus a picornavirus. The virus reproduces in the liver, damage to which can result in jaundice, and passes into the faeces through the bile duct.

Outbreaks of hepatitis A associated with sludge have involved one among sewage sludge spreaders and one transmitted by milk infected from process water contaminated by cesspool sludge (Havelaar & Block 1986).

Hepatitis B virus is not related to HAV. Hepatitis B virus is carried in blood and blood products, saliva or semen, which undoubtedly find their way into sewage in small quantities, but the so-called 'Dane particles' associated with hepatitis B virus particles are not demonstrable in faeces (Ketchum 1988). Hepatitis B virus cause serum hepatitis, which can be very mild but can also be fatal. Structurally, the hepatitis B virus could hardly be more different from the hepatitis A virus, as it has an envelope and a nucleocapsid containing DNA which is partly double-stranded. Other hepatitis viruses include, non-A, non-B hepatitis virus, which has properties in common with both HAV and hepatitis B virus, as it can be transmitted from person to person and cause water-borne epidemics with a very high proportion of fatalities (Abel 1989, Rao & Melnick 1986).

8.7 VIRUSES IN WATER

The ability of enteric viruses to cause water-borne disease in the form of human gastroenteritis is attributed to their low infectious dose and their survival in aquatic environments for prolonged periods. A survey of virus survival data (Kutz & Gerba 1988) showed that temperature is the principal factor in virus inactivation rates, although its precise significance varies according to the type of virus and the source of water. Virus survival in water is also affected by pH, light, organic content and the presence of micro-organisms. The overall average rate of reported virus inactivation at ambient temperatures is about 0.5 log-decade per day, decreasing in one day to about a third of the original number. Viruses were reported as being inactivated most rapidly in tap water, and coliphage in groundwater was the most slowly inactivated virus.

The effect of temperature on virus inactivation can be expressed as a linear correlation in terms of L_T (log-decades per day at temperature T°C) against temperature T(°C) of the form

$$L_T = L_0 + m_v . T \tag{8.1}$$

where L_0 is the value of L_T extrapolated to 0°C and m_v is the temperature coefficient of virus inactivation-rate in log-decades per °C-day. The overall correlation for all viruses reported was for a value of L_0 of 0.08718 log-decades/day and a temperature coefficient of −0.0284 log-decades/°C-day, so that

$$L_T = 0.08718 + 0.0284(T) \tag{8.2}$$

for all virus and water types. The overall temperature coefficient is similar to that for echovirus in impounded water and rotavirus in river water.

A similar approach was taken on a more detailed bases by Hurst (1988). A survey of published reports of survival of enteric viruses, either enteroviruses or rotaviruses, in the water of natural streams, rivers, lakes or ponds identified pH, chloride, total organic carbon, hardness, temperature, turbidity and sunlight as significant factors. Inactivation rates obtained from these surveys were correlated with temperature in the form of equation (8.1). For both rotaviruses and enteroviruses, the values of L_0 are about zero, and the temperature coefficients are 0.0530 and 0.0124 log decades/°C-day for enteroviruses and rotaviruses respectively, so that rotaviruses appear to be more stable in surface freshwater than enteroviruses. The combined effects of turbidity and sunlight show that turbidity inhibits the penetration of ultra-violet light into the water. In the presence of sunlight, the inactivation rate of poliovirus in river-water is reduced from 2.38 log-decades/day in low-turbidity water to 1.33 log-decades/day in high-turbidity water. In the absence of sunlight, the inactivation rates are much the same at 0.81 and 0.71 log-decades/day in high- and low-turbidity river-water respectively.

8.8 VIRUSES IN DISPOSAL OF SEWAGE AND SLUDGE TO SEA

Large quantities of domestic sewage and sludge are discharged or dumped directly into the marine environment. Discharge of sewage to sea through an outfall, usually after screening and maceration, is frequently practised as an alternative to conventional sewage treatment. Dilution by well-aerated seawater, enhanced by the salt content, strong tidal currents and the effect of ultra-violet radiation, are assumed to give the equivalent of 99% treatment in a conventional treatment plant. However, the evidence that pathogenic organisms, and viruses in particular, can survive for a very long time in the marine environment is substantial.

About 1½ million tons of sewage per day are discharged into the sea from the UK, and it has been estimated that the daily discharge of municipal sewage into the coastal waters of the United States amounts to 8 billion US gallons, about 30 million tons, of which only half receives secondary treatment. (Gerba & Goyal 1988). These wastes contain viruses which are pathogenic to man and can cause diseases such as hepatitis, gastroenteritis, meningitis, fever, rash, conjunctivitis. Infectious hepatitis and viral gastroenteritis can be caught by consuming raw or partially cooked shellfish, and aquatic recreational activities, such as swimming, can also be involved in the transmission of enteric viruses.

The risks from pathogenic viruses associated with the marine disposal of sewage sludge have been evaluated by Gerba and Goyal (1988). Investigation of marine sludge-dumping sites have shown that faecal indicator bacteria and viral pathogens occur in the surface water and accumulate in the sediments. At temperatures below 10°C, enteric viruses can survive for several months, and enteric viruses have been detected in sediments 18 months after cessation of sludge dumping. Gastroenteritis and ear infections may result from swimming in water contaminated with sewage, and the surf from contaminated ocean water creates viral aerosols in which the concentration of organisms in the aerosol can be 50 to 1000 times greater than in the water. Winds can transport marine micro-organisms as far as 100 miles from the ocean.

8.8.1 Viruses in shellfish

Filter-feeding shellfish tend to concentrate bacteria and viruses whose concentration in the shellfish can be many times that in the surrounding water. Outbreaks of viral disease can the occur by direct consumption of virus-contaminated shellfish, or by viruses entering the food chain through consumption of shellfish faeces by polycheate worms, which are then consumed by crabs. Even when shellfish are not harvested near sewage discharge points, mobile creatures such as crabs may move into nominally safe harvesting areas.

Of the hundred or so known enteric viruses, only a few are known to be transmitted by shellfish, although this may be owing to difficulties in detecting and recognizing viruses and the diseases caused by them (Gerba & Goyal 1988). Viruses reported as causes of illness associated with shellfish are astrovirus, hepatitis A, non-A, non-B hepatitis, Norwalk virus, Snow Mountain agent and small round viruses.

Mussels, the edible shelfish, are commonly used in assessing marine pollution, as an aquatic analogue of the miner's canary. HAV is one of several viruses in whose transmission to man shellfish are key vectors. HAV and cell-culturable viruses were determined in oysters, *Crassostrea angulata,* and mussels, *Mytilus edulis*, kept in cages near the outfall of a sewage treatment plant and compared with levels in samples of seawater and of effluent, at fortnightly intervals over a period of six months (Pietri *et al.* 1988). While all the effluent samples showed the presence of some enterovirus, with one sample as high as 326 pfu/100 ml, only three effluent samples were positive for HAV, one mussel sample was positive for adenovirus, and no seawater or oyster sample was positive for HAV, adenovirus or enterovirus.

An investigation of the effects of seawater contamination level on the rate of self-purification in mussels, *Mytilus edulis,* from indicator organisms, *Escherichia coli,* faecal streptococci and coliphage, showed that current commercial purification procedures for shellfish do not completely remove the indicator virus (Mesquita 1988). The mussels were kept on PVC trays in 150-l PVC tanks through which seawater at 2.5% salinity was circulated. After acclimatizing the shellfish, the seawater was contaminated with the indicator organisms as a culture or in settled sewage. The mussels were exposed either to a high dose for a short time, 10^5 to 10^7 organisms/litre as cfu or pfu for 30 minutes, or to a low dose for a long time, 10^3/litre for 24 hours. The mussels were then transferred to a similar cleaning tank, through which filtered seawater was circulated and treated with ultra-violet radiation at a rate of one tank volume per hour. These procedures are based on removing *Escherichia coli* within 48 hours, but left residual coliphage indicator virus present after 72 hours.

In the high-dose short-time tests, the indicator coliphage was removed from the mussels somewhat more slowly than the faecal streptococci, which were removed more slowly than the *Escherichia coli.* The mussels contained significant quantities, well above guideline values, of all three indicator organisms, after 72 hours' cleaning.

In the low-dose long-time tests, the level of indicator virus remained much the same throughout 96 hours of purification, although the *Escherichia coli* and faecal streptococci reached guideline levels after 30 hours and 72 hours respectively

The results suggest that bacteria, and particularly *Escherichia coli,* are not good indicators of viral contamination, although the effects on the cleaning procedure on coliphage may be rather different from that on human enterovirus.

8.8.2 Viruses in bathing waters

The survival of viruses in seawater has been well investigated (Gerba & Goyal 1988). The principal factor is temperature, and other factors affecting virus survival include salinity, microbial antagonism, sunlight and association with solids and sediments.

The results of an intensive survey over three years of the incidence of enteroviruses around the Welsh coast have been reported (Merrett *et al.* 1989). At 48 sea-bathing locations round the coast of Wales in 1986–88, 168 out of 623 samples taken (27%) were positive for enteroviruses, with virus counts in the range 0.1 to 2.3 pfu/l. In general, the incidence of viruses was greatest in locations close to areas of high population density, although viruses were found in waters with little sewage input or evidence of contamination by sewage from the *Escherichia coli*–total coliform count. However, 46% of samples complying with the mandatory *E. coli*–coliform standard failed to comply with the enterovirus standard, and there was no correlation between the presence of enteroviruses and the presence of coliform organisms. There is clearly a need for microbiological indicators of acceptability of quality of bathing waters more effective than coliform organisms.

In a survey of the virological quality of bathing waters at 148 sea-bathing locations in England in 1988, nearly 30% of samples taken (96 out of 330) were virus-positive (Morris & Cox 1989). Of these, 23% (77) showed enterovirus levels of 1 pfu/l or less, while 6% (19) were above this level. The highest value found was over 20 pfu/l. In a correlation of bacteriological and virological findings, of 87 locations, 16 (18%) complies and 40 (46%) failed to comply with EC bathing-water standards both bacteriologically and virologically, while 23 (26%) complied virologically but failed bacteriologically, while 23 (26%) complied virologically but failed bacteriologically. Although the likelihood of viral contamination increases as the level of bacterial indicators increases, viruses can be detected when no indicator bacteria are present.

8.9 VIRUSES IN SOILS

The persistence of viruses in soil is relevant when sludges and other wastes are applied to land. Virus survival in an environment depends on temperature, microbial activity and the type of virus, and in soil, also on saturation, soil chemistry and attachment to particles.

A study of the survival and transport of HAV, poliovirus 1 and echovirus 1 in soils showed that, of the three viruses tested, poliovirus were adsorbed most extensively and echovirus least (Sobsey *et al.* 1986). Clay soils were the most effective in adsorbing viruses and sandy soils the least, and in general the level of pollution of the water containing the viruses had little effect on adsorption. Sobsey *et al.* concluded that poliovirus 1 and echovirus 1 are unsuitable as models for predicting the adsorption, survival and transport of HAV in soils.

8.9.1 Temperature

Temperature is the predominant factor determining enterovirus inactivation rate. Viruses are inactivated by heat and survive longer at lower temperatures. Data on virus inactivation on soils cited in their review (Jørgensen & Lund 1986) showed that the virus inactivation rate was strongly affected by temperature. In studies of virus survival on soils over a protracted period, only the higher temperatures in the exposure range significantly affected virus survival. Other inactivating factors were sludge drying and exposure to ultra-violet light. Coxsackievirus in sewage sludge applied to land decreased to undetectable levels from an initial level of 10^9 virus per litre in 23 weeks in winter conditions with temperatures ranging from 0 to 10°C. The count of poliovirus 1 in sludge of soil decreased from 1000 virus per gram of soil to undetectable levels in 96 days in winter (−14 to +27°C) but in only 11 days in summer (+15 to 34°C). Other results cited showed that atmospheric humidity also affects virus survival, with levels of poliovirus 1 decreased from 10^7 virus per gram to undetectable levels in 35 days in warm, humid conditions, but in only 21 days in warm, dry conditions. Over a holding period of 28 days, poliovirus type 1 underwent a 57% reduction at 7°C but 97% at 20°C (Green 1976 cited by Snowdon *et al.* 1989a). Degradation of both porcine and human enteroviruses is principally temperature-dependent and the extent of inactivation is greater in the presence of oxygen (Lund *et al.* 1984 cited by Snowdon *et al.* 1989a). The effect of temperature for different periods of exposure is summarized in Fig. 8.1.

8.9.2 Ionizing radiation

The dose of ionizing radiation required to produce inactivation of a virus population is about 150 to 200 krad, as would be expected of such small targets, although bacteriophages appear to be more sensitive and require only 60 krad (White 1984).

8.9.3 Predation

Bacteria are known predators of enteric viruses, and may secrete inhibitory metabolites or utilize the virus capsid as a nutrient source. Poliovirus 1 is susceptible to some microbial proteolytic enzymes (Herrman *et al.* 1974 cited by Snowdon *et al.* 1989a). As microbial activity is strongly dependent on temperature and oxygen availability, these factors will thus also affect virus persistence, so that virus inactivation in soils due to microbial activity may occur only in aerobic conditions and at moderately high temperatures. Loss of virus infectivity in activated sludge due to microbial activity has been reported (Ward 1981 cited by Snowdon *et al.* 1989a). Two enteroviruses were inactivated more rapidly in a lake than in sterile lake water (Herrman *et al.* cited by Snowdon *et al.* 1989a) and more rapid virus inactivation was found in non-sterile waste-water (Sobsey *et al.* 1980), although survival of poliovirus 1 was much the same in sterile and non-sterile sand columns (Green 1976 cited by Snowdon *et al.* 1989a).

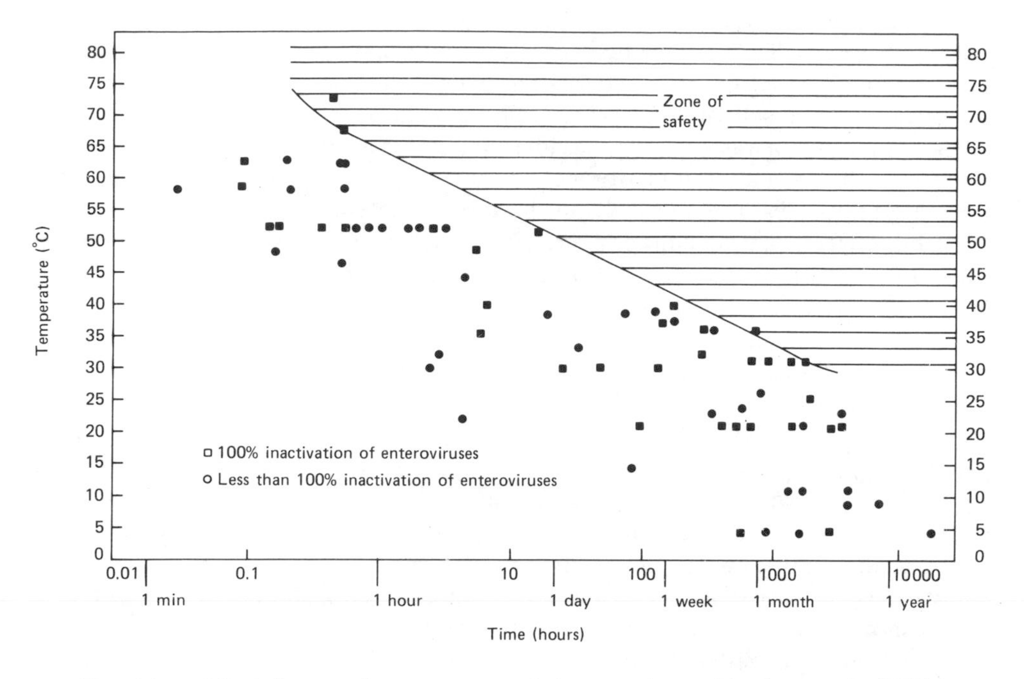

Fig. 8.1 — The influence of temperature and time on viruses. Feachem *et al.* (1983). Reproduced by kind permission of the World Bank.

8.9.4 Virus survival on crops and soil

The survival of poliovirus on crops and soil has been summarized from a survey of published data produced for the US EPA by Kowal in 1985 (cited by Schwartzbrod *et al.* 1987). The results are wide-ranging, but the survival of poliovirus on soil appears to depend more on the temperature than on the type of soil, which suggests that the effect of moisture is less important for a naked virus than it is, for example, for helminth eggs. On sandy loam the survival time increased from 12 days at 37°C to over six months at 4°C. Poliovirus survived on sandy soil in moist conditions for 91 days and less than 77 days in dry conditions at unspecified temperatures. On moisture-retaining clay soil, poliovirus survival ranged from 11 to 96 days at temperatures between 14°C and 33°C, and less than 161 days with rainfall. On crops, poliovirus survival was reported as 20 days on radishes at cool temperatures between 5 and 10°C and from a week to five weeks on tomatoes. Kowal (1982 cited by Snowdon *et al.* 1989a) also reported that survival of enteroviruses in soils can range from three to 170 days, depending on temperature and soil-type, and from one to 23 days on crops.

8.10 VIRUSES IN SEPTIC TANKS

The survival of viruses in septic tanks was reviewed and investigated by Snowdon *et al.* (1989a,b). Virus particles have very low settling velocities, and septic tanks will remove only viruses attached to solid particles with a realistic settling velocity. It has been determined that half the viruses entering septic tanks are associated with solids, but may be released into the supernatant when sedimented solids are disturbed. Significant numbers of viruses pass through septic tanks.

Virus survival in wastes is very variable but is generally days or weeks, rather than minutes or hours. Laboratory storage of enteroviruses at −18°C gives about 90% decrease in infectivity in a year. Thus duration of detention is a key parameter in preventing virus transmission, and any treatment involving an increased detention time, increased microbial population and increased temperatures, must be more attractive in terms of virus inactivation.

Within septic tank sludge, virus inactivation is affected by temperature, pH and oxygen availability, as viruses are inactivated more slowly in anaerobic conditions than when aerated, so that aeration gives faster inactivation (Lund *et al.* 1983, 1984 cited by Snowdon *et al.* 1989a). Virus removal by death or inactivation occurs much more slowly in septic tanks than removal by sedimentation within the septic tank or by displacement with the septic tank effluent.

In order to assess the advisability of discharging animal waste mixed with septic-tank effluent on to land, the inactivation of poliovirus 1 was studied in a laboratory simulation of storage of mixed animal waste and septic tank effluent at 5°C, 15°C and 25°C (Snowdon *et al.* 1989b). Times for 90% reduction of titre were 102, 71 and 22 days respectively in buffered saline; 80, 72 and 19 days in septic-tank effluent, but 19, 20 and 17 days in septic-tank effluent mixed with dairy manure slurry. This indicates that inactivation of the virus in septic-tank effluent is dramatically accelerated by mixing with animal slurry, particularly at low temperatures. This was attributed to bacteria, organic acids and solids present in the slurry.

8.11 VIRUSES IN SLUDGES

The fate of human enteric viruses, rotavirus and enterovirus, in sewage in primary sedimentation was reported by Bosch *et al.* (1986). The levels of enteric viruses, rotavirus and enterovirus, were both several thousand counts per litre, although no correlation was found between the levels of the two viruses. The effluent supernatants from both mechanical sedimentation and lime conditioning contained about ten virus counts per litre, which supports the view that virus particles in sewage are attached to sludge particles rather than in free suspension. The mechanically sedimented primary sludge gave virus counts of about 100 000 per kilogram, while viruses were in most cases not detected in the lime-treated sludge. It was concluded that viruses were inactivated by the high pH generated in lime treatment.

Schwartzbrod *et al.* (1987) summarized reported counts of enteric viruses isolated from sludges. Levels in primary sludges ranged from 132 to 140 000 pfu/l. Secondary sludges always had lower virus counts, ranging between 6 and 8210 pfu/l. Virus counts in digested sludges were always lower than in primary sludges, ranging from 4 to 4100 pfu/l. At first sight, it appears somewhat surprising that digested sludge has much the same levels of virus as secondary sludge, but it should be borne in mind that the digester feeds were probably mixtures of primary and secondary sludges, and so started with higher virus counts.

Analysis of 57 samples of different types of sludge from a wastewater treatment plant in southern France, some after five years' storage, showed that surplus activated sludge, and mixtures of primary sludge and surplus activated sludge all contained enteroviruses (Schwartzbrod *et al.* 1987). Only 44% of samples of sludge treated by mesophilic anaerobic digestion, at 35°C with a retention time of 15 to 20 days, and 10% of samples of sludge treated by chemical conditioning, with 20 to 25% lime plus 3.5 to 6% ferric chloride, and de-watering were positive for enteroviruses, and no viruses were found in sludge samples stored for a year or more. The primary sludge mixtures showed enterovirus counts of 645 to 3200 pfu/l, with a mean value of 1700 pfu/l, and the counts for surplus activated sludge were generally smaller by a factor of 10.

It is recognized that virus survival increases with decreasing temperature, and in cold and temperate climates, near-freezing conditions may persist for extended periods (Berg *et al.* 1988). The survival of enteroviruses at 5°C was tested in sludges from extended-aeration and oxidation-ditch sewage-treatment plants within the pH range 6.3 to 7.4. The pH range is important, as it affects the level of virucidal unionized ammonia, and also the virucidal activity of certain detergents. Both the extended-aeration plant and oxidation-ditch operate at very low sludge loadings ('food-to-micro-organism ratio') with mean hydraulic residence times of over 24 hours.

At 5°C, enteroviruses survived with undiminished numbers as long as 38 days in extended-aeration sludges and up to 17 days in oxidation-ditch sludges. However, large numbers of enteroviruses survived in sludges treated at pH 3.5 for one hour, which is difficult to explain, because anionic detergents commonly found in sewage, such as sodium dodecyl sulphate, are highly virucidal at low pH levels.

8.12 VIRUSES IN SEWAGE TREATMENT

The removal of viruses during sewage treatment has been reviewed by Rao *et al.* (1988), who cite counts of human rotavirus as ranging from one to several hundred per litre. Viruses in sewage are associated with the solids fraction, so that the major proportion, typically 65%, are removed by primary settling with a detention period of one to three hours. Activated sludge treatment processes were considered to be the most effective method for removing enteroviruses from sewage (Rao *et al.* 1988).

8.12.1 Activated sludge

Results cited by Rao *et al.* (1988) showed that laboratory-scale continuous-flow activated-sludge systems gave 90% removal of poliovirus and greater than 99% removal of coxsackie A9 virus. Similarly good performances have been obtained in plant-scale operations, with 90 to 99% enterovirus removal in one 18 000 m^3/day activated sludge plant and 90% enterovirus removal in a 10MGD (36 000 m^3/day) contact stabilization plant. The effect of different types of viruses was indicated in one plant which gave 93% removal of enteroviruses and 85% adenovirus removal, but only 28% removal of reoviruses. This suggests that enteroviruses may be more susceptible to removal by activated sludge treatment than other viruses (Rao *et al.* 1987, 1988).

A study of rotavirus removal in a sewage works treating 1.5 million gallons/day (5500 m^3/day) showed that primary settling removed 44 to 55% of indigenous rotavirus. Rotavirus removal, after activated-sludge treatment and primary settling, was 78 to 91% in the unchlorinated effluent, increased to 93 to 99% removal after chlorination of the final effluent (Rao *et al.* 1987, 1988). Based on figures from seeded viruses, it was likely that 92% of rotavirus is removed in activated sludge. Rotavirus counts varied seasonally in the range 40 to 500 per litre in raw sewage.

Activated sludge treatment is known to reduce concentrations of detectable enteric pathogens (Knowlton & Ward 1987). The removal of virus particles is generally believed to be due to their attachment to and removal with sludge particles, but evidence is now accumulating that viruses are inactivated by substances in the mixed liquor in the activated sludge process. Loss of viral infectivity observed in mixed-liquor suspended solids (MLSS), wastewater sludge, fresh water and seawater has been associated with such factors as ammonia, ionic detergents, light, heat and other micro-organisms. When seeded into MLSS from a contact-stabilization sewage-treatment plant dealing with mainly (over 90%) domestic sewage, poliovirus 1 and rotavirus SA-11, as model enteric viruses, were irreversibly inactivated and released RNA after incubation for 48 hours at 26°C, while echovirus 12, coxsackievirus A13 and reovirus 3 lost infectivity.

Procedures inactivating or removing micro-organisms in MLSS consistently removed virucidal activity, so that virus inactivation appeared to be micro-organism-related. Virucidal agents in MLSS are evidently active only while associated with micro-organisms themselves and/or are short-lived and must be continuously generated, because virucidal agents could not be separated from viable micro-organisms in MLSS. As proteolytic enzymes fit these descriptions, it was suggested that seeding sludge with organisms secreting appropriate proteolytic enzymes could increase the capacity of sludge solids to reduce the concentration of infectious viruses in sewage.

The results of batch experiments on virus removal by activated sludge treatment at different sludge loadings were reported by Shimohara *et al.* (1985). Counts of poliovirus 1 were reduced by 99% with sludge loadings between 0.2 and 0.4 kg(BOD)/kg (sludge solids)-day and sludge solids concentrations between 1 and 2 kg (sludge solids)/m^3. The minimum sludge concentration giving 99% virus removal was found to be 0.8 kg/m^3. In general, operating conditions giving good BOD removal also gave good virus removal.

8.12.2 Sewage effluents

Following recurring outbreaks of gastroenteritis in the very densely populated tropical island of Puerto Rico, a survey of ten percolator and five activated-sludge sewage-treatment plants showed that all 15 discharged effluents upstream of intakes to water-treatment plants (Dahling *et al.* 1989). Virus counts were generally several hundred pfu's per litre in the incoming sewage, exceeding 10^5 pfu/l at several plants. In the treated effluent, virus counts were generally several hundred pfu's per litre in the treated effluent, and even 24 000 pfu/l in the effluent discharged from one percolator plant, indicating defective sewage-treatment processes. Most (80%) of the enterovirus isolates were identified as coxsackievirus B5, usually associated with the severe clinical illnesses meningitis and encephalitis.

8.12.3 Maturation ponds

Studies of the disinfection efficiency of maturation ponds indicated that they are effective in treating poor-quality effluent from a percolator and in removing viruses and parasite ova, but can suffer from hydraulic short-circuiting as a result of the design of the pond and thermal stratification within the water (Macdonald & Ernst 1987).

8.13 VIRUSES IN SLUDGE TREATMENTS

8.13.1 Mesophilic anaerobic digestion

Monitoring the inactivation of a human rotavirus, a coxsackievirus B5 and a temperature-resistant bovine parvovirus during sludge-treatment processes indicated that all three viruses were inactivated to only a minor extent during conventional mesophilic anaerobic digestion at 35 to 36°C (Spillman *et al.* 1987). Over a week under these conditions, inactivation-rate coefficients of 0.314, 0.475 and 0.944 log-decade per day respectively were obtained.

The effect on virus survival of conventional single-stage pilot-scale mesophilic anaerobic digestion and two-stage anaerobic digestion processes, with and without sludge pasteurization before and/or after digestion, has been reported (Wekerle *et al.* 1985, Leuze *et al.* 1985). Three naked viruses were tested, a bovine DNA parvovirus, a bovine RNA enterovirus and a reovirus. The reovirus and bovine enterovirus were completely inactivated by single-stage digestion at 33°C with a retention time of 20 days. They were also completely inactivated by two-stage digestion, using anaerobic pretreatment at 33°C or 20°C with a two-day retention time followed by digestion at 33°C with an eight-day retention time, both with and without pre-pasteurization. However, none of the processes tested gave complete inactivation of the bovine parvovirus, even including pasteurization of the digested sludge. The possible cause of reovirus inactivation was attributed to detergents present in the sludge.

8.13.2 Thermophilic anaerobic digestion

In further work by Wekerle's group, (Saier *et al.* 1985), thermophilic anaerobic digestion (55°C) and subsequent mesophilic digestion of sludge on the survival of viruses without and with pasteurization of the digested sludge was investigated. The effect of two-stage digestion, comprising thermophilic anaerobic digestion (three days at 55°C) followed by mesophilic digestion (12 days at 33°C), on two RNA viruses (reovirus 1 and bovine enterovirus (BEV) ECBO-LCR-4) and a DNA virus (bovine parvovirus (Haden)), were compared with that of mesophilic anaerobic digestion (20 days at 33°C). Reovirus was not isolated from sludge treated by either process operating at pH levels between 7.24 and 7.35, nor bovine enterovirus from either process operating between pH 7.21 and 7.28. Parvovirus, however, was detected in samples from sludge treated by both processes, although parvovirus in sludge samples was destroyed by pasteurization for 30 minutes at 70°C.

Spillman *et al.* (1987) found that anaerobic thermophilic digestion at 54 to 56°C gave rapid inactivation, with rate coefficients greater than 8.5 log-decades/*hour* for rotavirus, greater than 0.93 log-decades/*minute* for coxsackievirus B5 and 0.213 log-decades/*hour* for parvovirus measured over periods of several hours.

8.13.3 Thermophilic aerobic digestion

Spillman *et al.* (1987) also found that aerobic thermophilic digestion at 60 to 61°C rapidly inactivated rotavirus, with a rate coefficient of 0.75 log-decades/minute, coxsackievirus B5, with a rate coefficient greater than 1.67 log-decades/minute, and parvovirus at 0.353 log-decades/hour over periods of several hours. Thermal inactivation predominated at temperatures above 54°C, and pasteurization at 70°C for 30 minutes gave an inactivation of 0.72 log-decades with parvovirus in that time. Thus the rotavirus and the coxsackievirus were rapidly inactivated at 55°C, whereas the parvovirus is unusually thermostable. The rotavirus and enterovirus are resistant to chemical factors such as ammonia and detergents, whereas the parvovirus is chemically labile. It was therefore suggested that parvovirus acts as a reliable indicator for assessing thermal inactivation, while rotavirus acts as an indicator for other inactivation processes. It was concluded that, while mesophilic anaerobic digestion should eliminate thermostable, chemically sensitive viruses, it should be preceded by heat treatment involving temperatures around 60°C to inactivate temperature-labile chemically resistant viruses.

Further reports of the effects of different sludge treatments on pathogens generally, including viruses, are given in chapter 9.

8.14 EFFECTS OF CHEMICAL TREATMENT ON VIRUSES

8.14.1 Chlorination

Reports of the isolation of HAV and other viruses from treated drinking water suggest that, under certain conditions, some viruses are able to survive treatment, and that HAV is more resistant to water chlorination processes than other enteroviruses and indicator bacteria. It has been reported that, although HAV was more sensitive to free chlorine than certain indicator viruses, HAV was relatively resistant to combined chlorine in tap water and sewage effluent (Grabow *et al.* 1984). The extent and rates of inactivation at 5°C of HAV and three other viruses, coxsackievirus B5 (CB5) and coliphages MS2 and φX174, by 500 mg/m^3 (0.5 mg/l) free chlorine at pH 6 to 10, and by 10 mg/l monochloramine at pH 8 were investigated by Sobsey *et al.* (1988).

Both coliphages were sensitive to free chlorine, and all four viruses were relatively resistant to monochloramine. HAV proved to be rapidly inactivated by free chlorine, but only slowly by monochloramine, compared with the other viruses. HAV and CB5 had similar resistances to monochloramine, but HAV was less resistant then CB5 to free chlorine. The coliphage MS2 was shown to provide a reasonably good model of HAV inactivation by free chlorine and monochloramine, and of CB5 by monochloramine. Coliphage MS2 does not model inactivation of CB5 by free chlorine well. It was also suggested that the inactivation efficiency may be affected by virus aggregation and by the effect of pH on virus conformation. It is known that poliovirus 1 and echovirus 1 can exist in at least two pH-dependent conformations, and there is some preliminary evidence that one type of HAV may also. For the same reason, pH may also affect virus infectivity.

8.14.2 Ozone

When chlorine is used as a disinfectant, there is a risk of formation of carcinogens such as trihalomethanes, and ozone is a promising alternative. Ozone has been shown to give rapid inactivation of several viruses including poliovirus -1, -2 and -3, vesicular stomatitis, encephalomyocarditis, adenoviruses, coxsackievirus and, with some strain variation, echovirus (Vaughn *et al.* 1987). Both human and simian types of rotavirus were inactivated rapidly by ozone at concentrations of 0.25 mg/l or greater in the pH range 6, 7 and 8 at 4°C, with the human type more sensitive to ozone. At the concentrations used, the virucidal effect of ozone on the rotaviruses was similar to that of chlorine.

9

Pathogens in sludge treatment

9.1 INTRODUCTION

In the UK, most sludge is used as organic fertilizer on agricultural land, which requires good management to avoid accumulation of undesirable pollutants in the soil (Hall 1989b).

Sludge treatments are aimed at reducing the volume as quickly as possible by removing water from the sludge solids, stabilizing the sludge by reducing the content of easily biodegradable organic nutrients, and disinfecting the sludge by inactivating pathogenic organisms. The principal methods of sludge treatment are de-watering followed by storage and anaerobic digestion. The latter may also include combined heat and power (CHP) energy recovery. Processes under development are composting, incineration and thermophilic aerobic digestion. Other methods under study are co-disposal in landfill, with an opportunity for methane recovery, and in forestry, with a reduced risk of contaminants entering the food chain (Hall 1989b).

The details of procedures for volume reduction and de-watering are beyond the scope of the present discussion. They are nevertheless important, as most sludge treatments cannot operate efficiently, or even at all, on dilute sludges.

To render waste sludge hygienic, all organisms which are pathogens and/or indicators of faecal contamination must be destroyed or permanently deactivated, as, in theory at least, survival of a single viable micro-organism can result in re-infection. When this has been achieved, reinfection with pathogens must be prevented.

Waste sludge is liable to putrescence, and conventional sludge stabilization involves anaerobic digestion for a protracted period of time. In some countries, Switzerland, for example, this treatment is no longer considered satisfactory, because mesophilic anaerobic digestion fails to achieve complete elimination of infectious agents, pathogenic protozoa, bacteria, viruses and parasite eggs present in the original raw sewage. Heat treatments, such as pasteurization or thermophilic aerobic pretreatment, provide a possible answer to this objection (Hamer 1989).

Bacteria are generally more easily inactivated than parasite ova or cysts, so that sludge treatments designed for inactivation of *Taenia saginata* or *Ascaris* ova or

Cryptosporidium oocysts should also give adequate destruction of bacteria. Bacterial spores, however, are the organisms which, apart from prions, are the most resistant to degradation known, such as those of *Clostridium perfringens* whose germination may actually be stimulated by heat treatment.

9.2 INDICATOR ORGANISMS

A comprehensive microbiological investigation of the suitability of potential indicator organisms was carried out on raw and treated sludges from conventional sewage treatment plants, and the effects of temperature and pH on selected indicator organisms were tested (Strauch 1989). The effectiveness of different sludge disinfection procedures was checked by microbiological examination of disinfected sludge from eight sewage-treatment plants in Germany using different disinfection procedures, pasteurization, aerobic thermophilic stabilization (ATSD), ATSD followed by mesophilic anaerobic digestion, treatment with slaked lime or quicklime, and composting in windrows and in closed reactors.

The microbiological tests covered total aerobic micro-organism count, enterobacteriaciae, coliforms (section 7.1.1), faecal streptococci (section 7.1.2), *Salmonella* (section 7.6), fRNA phages (section 8.3) and the percentage of sludge samples infected with parasite ova. The potential indicator organisms selected were enterobacteriaciae, salmonellae, faecal streptococci, coliforms and fRNA-phages.

The F-specific or fRNA phages are so called because they infect the host bacterium through the fertility-pilus, which is involved in the exchange of genetic material between bacteria. The F-pilus is found on *Escherichia coli* and other related bacteria, and is produced only at temperatures above 30°C. The fRNA phage is thus less likely to be produced by enterobacteria in a temperate environment, and so is a good candidate for use as an indicator organism, with resistance to treatment presumed to be similar to that of viruses generally.

9.2.1 Indicator results

The results showed that there were no seasonal differences in the total aerobic micro-organism count and that in the treated sludge was about a third (half a log-decade) of that in the raw sludge.

9.2.1.1 Bacteria

Enterobacteriaciae varied between 0 and 10^8 cfu/ml in both raw and treated sludge: in winter, the values in raw sludge were about ten times higher than in treated sludge, but in summer about ten times smaller in raw than in treated sludge. In treated sludge, the summer values were higher than the winter values. The enterobacteriaciae counts were generally ten times (1 log-decade) higher than the coliform counts.

The *Salmonella* counts gave unexpected results, as *Salmonellae* were found in both raw and treated sludge samples more frequently in winter than in summer. *Salmonellae* were found in 87.5% of raw sludge samples in winter and 79.2% in summer; in 79.2% of treated sludge samples in winter and in only 41.7% in summer. The average disinfection effect of sludge treatment was to reduce the *Salmonellae* by 99% (two log-decades) to levels of 10 to 100 cfu/ml.

Faecal streptococci were present in virtually all sludge samples at levels of 10^3 to 10^5 cfu/ml.

9.2.1.2 Virus
The fRNA-phage count was generally three to ten times higher (half to one log-decade) in the raw sludge than in the treated sludge.

9.2.1.3 Parasite ova
Parasite ova were found in winter in 70.8% of raw sludge samples and in 45.8% of treated samples; in summer, in 50% of raw sludge samples and in 54.2% of treated sludge samples. The infectivity of the eggs could not be evaluated.

9.2.2 Conclusions on indicator organisms
From these results, it was concluded that the total aerobic micro-organism count cannot be used as a reliable indicator of the state of hygiene of sludge. Although there appears to be a significant correlation between coliform and virus counts in raw sludge samples, this does not appear to apply to treated sludge samples. In addition, other investigators could find no coliforms in samples of groundwater, surface water, recreational water and drinking water in which viruses were present.

In raw sludge samples, there was a significant correlation between coliforms and *Salmonellae*. However, the coliform count has been found to be unreliable as an indicator of *Salmonellae* in other contexts, notably in a recent *Salmonella* epidemic in California. A correlation found between enteroviruses and *Salmonellae* suggested that *Salmonellae* could be used as a virus indicator. The use of fRNA phages as virus indicators is dubious, because they have been found in raw sludges where enteroviruses were undetectable.

Overall, from this investigation, together with results from other similar work reported in the literature, it was concluded that several micro-organisms in raw and treated sludges can be used as indicators of other organisms, but the true extent of sludge contamination can ultimately be assessed only by direct determination of the relevant organisms.

9.2.3 Virus results
Raw sludge samples generally had higher enterovirus counts in summer than in winter, while the opposite was true of treated sludge samples. Conventional sludge treatment results in 42.6% reduction in virus count in winter and a remarkable 82.7% reduction in winter.

No correlations were found between fRNA phages and either enterobacteriacae or enteroviruses, in either raw or treated sludge samples. In treated sludges, there were correlations between enteroviruses and both faecal streptococci and *Salmonellae*.

9.2.4 Effects of temperature
The survival of organisms in raw sludge at pH 3 is summarized as follows.

- The **total aerobic micro-organism count** was reduced by 1000-fold (3 log-decades) after five hours at 60°C, the highest temperature employed.
- **Enterobacteriaciae** were not detectable 48 hours at 50°C, four hours at 55°C or two hours at 60°C.

- *Salmonellae* were not detected after 24 hours at 50°C, four hours at 55°C and 45 minutes at 60°C.
- *Ascaris* eggs were not detectable after four hours at 50°C, 30 minutes at 55°C and 15 minutes at 60°C.
- **Enteroviruses** were inactivated after 60 hours at 50°C, ten hours at 55°C and 15 minutes at 60°C.

9.2.5 Effects of lime treatment of sludge

Two levels of dosing raw sludge with slaked lime, $Ca(OH)_2$, were tested, 7 and 10 kg/m^3. Enterobacteriae were not detected after 24 hours at either level, *Salmonellae* after 48 hours and five hours, fRNA phages after one hour and 15 minutes, respectively, and enteroviruses were inactivated after ten hours and three hours respectively.

9.2.6 Reliability of sludge disinfection methods

Tests of the efficacy of sludge disinfection in eight sewage-treatment plants showed that five of the plants (thermal conditioning, pasteurization of digested sludge, reactor composting, ^{60}Co irradiation) had unacceptably high levels of enterobacteriaciae and *Salmonellae* in the treated sludge, and two of the other three (aerobic thermophilic digestion, liming) had treated sludge samples positive for parasites. Several also had enterovirus counts over 200 pfu/l (two thermal conditioning plants, a different aerobic thermophilic digestion plant and the pasteurization plant cited previously).

9.2.7 Conclusions

Taking other results into account, the laboratory tests showed that *Salmonellae* were unexpectedly resistant to high temperatures and high pH levels. Although the tests used unusually high initial concentrations of 10^4 to 10^7 cfu/ml, the results are borne out by the high residual *Salmonellae* counts in some of the treated sludge samples. This underlines the importance of testing for *Salmonellae* in assessing sludge disinfection procedures. Faecal streptococci cannot be used as hygiene indicators because their susceptibility to sludge treatments differs considerably from that of *Salmonellae*. High counts of faecal streptococci are not necessarily associated with high *Salmonella* counts, and low faecal streptococci counts do not necessarily signify low *Salmonella* counts. Faecal streptococci do, however, appear to give a reliable indication of virus contamination. It was concluded that sewage can be considered safe when it is free of *Salmonella*, has less than 1000 enterobacteriae per gram and viruses less than 200 pfu/l.

9.3 PASTEURIZATION

In pasteurization, the sludge is heated to a high temperature for a relatively short time, for example 70°C for 30 minutes. The process has been developed for elimination of *Salmonellae*, but it does not affect the tendency of the sludge to putresce. Sludge is, after all, a richly nutrient material. After storage, pasteurized sludge may contain higher numbers of enterobacteria than raw sewage, as the pasteurized sludge has been found susceptible to regrowth of enterobacteria, including *Salmonellae*. Re-growth is a problem only with bacteria, as parasite eggs, protozoan cysts and viruses can grow only in their animal or human host. The suceptibility of pasteurized sludge to re-growth of enterobacteria is thought to be due to the removal of competing microflora, as re-growth does not occur if pasteurized sludge is fed into an anaerobic digester (Havelaar 1984). Sludge pasteurization should include sufficient comminution to give no sludge particles larger than 5 mm (Strauch 1989), in order that the whole of the sludge solids should reach the required temperature.

In a pilot-scale system (Philipp 1981), coliform levels were 10^7 to 10^8 per millilitre in the raw sludge. The coliform counts were 10^4 to 10^5/ml after mesophilic anaerobic digestion, falling to 0 to 100/ml after subsequent pasteurization, but then rising to 10^6 to 10^7/ml after open-air storage.

The coliform count after pasteurization of raw sludge was also 0 to 100/ml, rising to 10^3 to 10^4/ml after digestion, but with no further increase on storage. Levels of *Salmonella* were low throughout the process, and deliberately introduced *Salmonella senftenberg* were killed by pasteurization.

Investigation of pasteurization conditions indicated that, although 60°C for 30 minutes should be sufficient, 65 to 70°C is needed to allow for the reduction in the effectiveness of heat treatment by inhomogeneity and aggregates of solid matter. Pasteurization was concluded to be highly effective for killing vegetative bacterial cells, parasite ova and some viruses, but certain viruses (e.g. 'bovine parvo') and bacterial spores are thought to survive. Ova of *Ascaris lumbricoides* are killed by 30 minutes at 60, 65 or 70°C (Strauch & Berg 1980).

9.3.1 Pasteurization with submerged combustion

A system for pasteurization by submerged combustion together with anaerobic digestion has been reported (Kidson & Ray 1984). Biogas produced by anaerobic digestion was used to pre-heat and pasteurize the digester feed in a submerged combustion system. Residence time in the burner tank was three hours, achieving a temperature between 38° and 55°C, and the sludge was cooled to 35°C before feeding to the digester.

Preliminary tests in peptone showed that the die-off rate of *Salmonella* was dependent on both pH and temperature, but at 44°C, 99.9% reduction occurred in less than 20 minutes over a range of pH values. In sludge at 55°C and pH 5.9, 99.999% reduction of *Salmonella* was obtained in one hour.

With ova of *Taenia saginata*, 90% loss of viability was obtained in 5.2 hours at 50°C and in 2.9 hours at 55°C. The composition of the medium and its pH was not specified.

9.3.2 Pasteurization with microwave heating

Microwave heating is commonly used in food preparation, and is thus worthy of investigation as a potential sludge treatment technique. Results of microwave treatment of liquid manure as an alternative pasteurization process for the disinfection of sewage sludge have been reported (Niederwöhrmeier *et al.* 1985). As mentioned in section 3.4.2, cysts of the sugar-beet nematode *Heterodera schachtii* in soil samples have been inactivated by microwave irradiation of 650 W for two minutes, and virus inactivation in water and in liquid manure has been reported.

Suspensions of four different species of bacteria, *Escherichia coli, Salmonella senftenberg, Erysipelothrix rhusiopathiae* and *Yersinia enterocolitica*, at 10^6 to 10^7 cells/ml in drinking water were subjected to heat treatment using a microwave flow heater. The temperature achieved increased with the residence time in the microwave field, increasing in an approximately linear fashion between 40°C with three seconds to 70°C with eight seconds. All species were inactivated by a residence time of 7.5 seconds in the 1 kW, 2450 MHz microwave field, giving temperatures up to 69°C. It was not, however, clear how long samples remained at the test temperature before incubation. Recent concern over inadequate cooking of food in microwave ovens has centred on the holding time following rapid heating.

The most temperature-sensitive organism was *Erysipelothrix rhusipathiae*, the most temperature-resistant was *Salmonella senftenberg*. None of the bacterial species tested was much affected by temperatures below 50°C, and all were inactivated at 69°C. The most temperature-resistance *Salmonella* sp. was then tested in untreated, unaerated liquid manure and the inactivation conditions were found to be similar to those in water.

9.3.3 High-pH pasteurization

A sludge pasteurization process using a combination of high pH and high temperature, called the **active sludge pasteurization process** has been described (Barnard *et al.* 1990). Sludge was first de-watered to give the highest solids content, using gravity thickening, while remaining pumpable, about 12% solids. The thickened sludge was then pumped through a continuous plug-flow system into which anhydrous ammonia was injected to raise the pH level to 11.5, followed by a contact time of 5 minutes, to allow the effects of the high pH and ammonia concentrations to destroy pathogens. Sufficient phosphoric acid is then injected into the sludge flow to neutralize the ammonia, with the heat of reaction increasing the sludge temperature to at least 65°C, followed by a mixing and contact time of two minutes. The pasteurized sludge is then cooled by heat exchange with the sludge feed.

The product comprises a liquid organic fertilizer containing 2.5% nitrogen and 4.7% phosphorus. In test runs, the reactor temperature could be controlled by the addition of ammonia and acid. Tests on the sludge product detected no faecal *E. coli, Salmonella, Clostridia* or anthrax organisms, nor eggs of *A. lumbricoides*.

High-pH heat treatment has been used successfully in disinfecting waste biomass from industrial fermentations using genetically engineered organisms, yeasts and bacteria, for use as fertilizer (Grüttner & Larsen 1990). The residual fermentation broth after product separation contains about 5% dry solids, of which about half is organic matter and half inorganic. The heavy metal content is about the same as in pig slurry and 10 to 50 times than in municipal sludge. The fermentation sludge is treated with lime to raise the pH above 11, heated to above 90°C using a combination of spiral and concentric heat exchangers and direct steam injection. The sludge is held at this temperature for at least one hour, cooled to less than 40°C and used as fertilizer. The treatment conditions were tested using wild-type organisms, as genetically engineered strains are considered to be much more sensitive to adverse environmental conditions. *Escherichia coli* was reduced from 10^9 cfu/ml to undetectable levels after ten minutes at 90°C with pH levels of 8.4 and 10; *Saccharomyces cerevisiae* from 10^7 cfu/ml to undetectable levels after ten minutes with pH 7 and

90°C or pH 10 and 80°C. Spores of *Bacillus subtilis* at pH 10 were reduced from 2.5×10^5 cfu/ml to undetectable levels after ten minutes at 100°C, but a few were still detected after 60 minutes at 90°C.

9.4 ANAEROBIC DIGESTION

Large-scale stabilization and disinfection of sewage sludge by two-stage anaerobic thermophilic and mesophilic digestion was investigated over a period of a year (Mitsdörffer *et al.* 1990). After thickening of the raw sludge, preliminary digestion is carried out at 55°C with a retention time of five days, followed by 15 days' mesophilic digestion at 35°C. The numbers of enterobacteriaciae were reduced from 10^8/ml in the raw sludge to about 10/ml after the first stage and after the second stage.

The method of feeding fresh sludge and withdrawal of digested sludge can affect the disinfection performance of a digester. To avoid infection of treated sludge with untreated raw sludge, it is safest to withdraw treated sludge from the digester before feeding in the raw sludge. However, it is usually convenient to feed fresh raw sludge and draw off the resulting overflow of treated sludge. There is then a risk that some of the raw sludge feed will pass straight through to the overflow untreated, thus infecting the treated sludge.

The influence of the sludge feeding pattern on digester performance was investigated using a laboratory-scale digester by Farrell *et al.* (1988). In laboratory-scale single-stage anaerobic digesters working on primary sewage sludge, changing the feeding pattern from 'draw down, then fill' (D/F) to 'fill, then draw down' (F/D) had a considerable effect on reduction of virus and faecal coliform counts, although little effect on volatile solids removal. Reductions on both bacterial and virus counts were greater using the D/F feeding pattern than with the F/D pattern, although the effect was more marked with bacteria than with viruses. With a digestion temperature of 35°C and a mean hydraulic retention time of 14 days, once-per-day D/F operation gave reductions of 2.1 to 2.5 log-decades for faecal coliform and faecal streptococcus, compared with 1 to 1.4 log-decades for F/D operation, and reduction in total coliforms of 2.7 to 2.9 log-decades for D/F operation compared with 1.2 to 1.34 log-decades for F/D operation. The corresponding figures for virus reduction were approximately 1 log-decade and 0.75 log-decade for D/F and F/D operation respectively.

Feeding patterns with full-scale digesters appear to be chosen as a matter of operational convenience, with near-continuous operation giving the advantage of more uniform gas evolution, or with the feed mixed in to cause an overflow into the next stage. Removal of digested sludge before feeding, which is commonly used in industrial fermentations, is apparently unusual in large-scale sludge digestion, although, from these results, it appears to give greater pathogen removal.

A mathematical model based on first-order kinetics was developed by Farrell *et al.* to describe the effects of feeding pattern and hydraulic residence time on microbial densities.

$$VC_D\exp(-k\theta_A)+\upsilon C_F=(V+\upsilon)C_{D'}\exp(k\theta_B) \tag{9.1}$$

where V is the volume of the digester contents of composition C_D before a volume υ of feed of composition C_F is added; $C_{D'}$ is the composition of the withdrawn digester product, θ_A is the time interval between withdrawal of product and next addition of feed and θ_B is the time interval between addition of feed and next withdrawal of product, and k is the first-order decay-rate coefficient for the species considered. This equation applies for both D/F and F/D patterns. Average values of the decay-rate coefficients at 35°C obtained using this model were 2.7 day^{-1} for faecal coliforms, 2.5 day^{-1} for faecal streptococci, 3.9 day^{-1} for total coliforms and 0.5 day^{-1} for viruses. The hydraulic residence time affects the net bacterial decay rate insofar as it affects the nutrient availability in the higher contents, but should not affect the virus decay rate.

9.5 AEROBIC THERMOPHILIC SLUDGE DIGESTION (ATSD)

In ATSD, heat is generated by the aerobic metabolism of the organisms in the sludge, and at higher temperatures, thermotolerant organisms metabolize more quickly, generating heat more rapidly and raising the temperature of the digested mass. The inactivation rate of pathogens in the sludge then increases as the temperature increases.

Microbial death is usually assumed to follow first-order kinetics, and the specific death-rate coefficients depend on both temperature and the previous treatment to which the material has been subjected. For example, the effect of temperature on the specific death-rate coefficient for *Escherichia coli*, an indicator organism for faecal contamination, follows the Arrhenius relation, giving a 150-fold increase between 54°C and 62°C. Later data suggest that this may even underestimate the effect. This relation cannot be extrapolated to temperatures below 50°C, at which *Escherichia coli* undergoes a transition from irreversible to reversible thermal damage. Investigation of the effect of temperature on *Klebsiella pneumoniae*, a mild pathogen occurring in sewage, in a continuous aerobic culture system with a mean residence time of 32 hours, showed that the metabolic activity at 55°C and even at 60°C, although low, was not zero (Mason & Hamer 1987). This suggested that completely mixed, continuous flow operation may not be suitable for achieving complete microbial inactivation in aerobic thermophilic digestion (Hamer 1989).

9.5.1 Standard processes

Aerobic thermophilic stabilization carried out with air gives temperatures of 40 to 60°C, or with high-purity oxygen, temperatures of over 60°C, as high as 80°C, depending on sludge solids content, retention time and aeration efficiency. Over three hours at 50°C are required to attain effective deactivation of *Salmonella, Ascaris suum* ova or bovine enterovirus. In full-scale plants, short-circuiting results in a significant proportion of the sludge being inadequately heat-treated, and the mixing characteristics of the reactor appear to be as important as its temperature–time characteristics. Aerobically digested sludge, after an average 20 days at about 49°C, still contained substantial amounts of enteric viruses and indicator bacteria, with counts of total coliforms reduced by 5.5 log-decades, faecal coliforms by 4.4, faecal streptococci by 3.4, and enteric viruses by more than 2.6 log-decades. In one system reported, however, digested sludge was withdrawn just *after* fresh sludge was fed in (Berg & Berman 1980). This involves the risk that the digested sludge withdrawn contains some of the virtually untreated fresh sludge previously added, as discussed in section 9.4. In studies of aerobic thermophilic digestion operating at 45 to 55°C with a retention time between 20 and 30 days, *Salmonellae* and enteric viruses were reduced to undetectable levels, faecal coliforms by 3.5 log-decades and faecal streptococci by 2.5 log-decades. The interval between feeding and withdrawal was 12 to 24 hours, and occasional viable parasite eggs were found in the digested sludge (Kabrick & Jewell 1982).

Data from the first UK full-scale, air-operated aerobic thermophilic digestion system (Morgan *et al.* 1984) showed a dramatic effect on the survival of *E. coli* and *Salmonella* spp, and indicated that in continuous systems, a retention time of eight days at temperatures above 50°C are needed for *Salmonella* removal. It was concluded that systems operating at lower temperatures would have to use discontinuous feeding to achieve adequate pathogen destruction. An ATSD system at a small sewage treatment works in the UK with a working volume of 17 m^3, an operating temperature of 55°C and a mean retention time of 11 days was reported as giving reductions of 3 log-decades in counts of *Escherichia coli* and enterobacteriaciae and 2 log-decades of faecal streptococci and *Clostridium perfringens* (Murray *et al.* 1990).

The suitability of aerobic thermophilic digestion, called 'liquid composting', for treatment of sludge from domestic wastewater from small communities was investigated using three full-scale operations (Kelly *et al.* 1990). In the three operations, digesters 10, 18 and 34 m^3 in volume were used, with each operational unit using two digesters in series, with average retention times of about 20, five and six days. Temperatures tested were 49–62°C in the 18-m^3 system, 67–70°C in the 10-m^3 system and 48–66°C in the 34-m^3 system. The effects of the treatment on coliforms, faecal streptococci and *Salmonella* were checked, although the *Salmonella* counts rarely exceeded 2/g in the feed. Coliform and faecal streptococci counts were generally reduced to less than two organisms per gramme from about 10^3 to 10^5/g of sludge feed. On the few occasions when the *Salmonella* counts in the feed reached several hundred per gramme, the treatment reduced the count to less than 2/g. It was concluded that the process produced satisfactorily hygienic treated sludge, except with temperatures less than 50°C and where the product of retention time and temperature was less than 130 day-°C, although the product of temperature and *log* time seems a more logical parameter to use. The odours from all reactors were considered acceptable.

9.5.2 Two-stage ATSD

In an investigation of the disinfection effects of single stage and two-stage aerobic thermophilic stabilization of liquid raw sludge, tests using *Salmonella senftenberg*, parvovirus (Haden) and eggs of *Ascaris suum* and assays of total aerobic microorganisms, coliforms and enterobacteriaciae, showed that inactivation of pathogens can be achieved by single-stage aerobic thermophilic digestion of liquid sludge under certain conditions (Strauch *et al.* 1985). These conditions generally involved residence times of three or four days at temperatures over 55°C, and maintenance of such in routine day-to-day operation proved to be difficult, with a continual risk of sludge being inadequately treated as a result of hydraulic short-circuits. It was therefore concluded that thermophilic aerobic digestion should be operated only as a two-stage system. In addition, coliforms were found to be inactivated before

Salmonella, so that coliforms cannot be considered to act as good indicator organisms for *Salmonella* removal in sludge treatment.

Strauch (1989) concluded that aerobic thermophilic digestion should operate as a two-stage process with two reactors in series, to avoid hydraulic short-circuits, with an overall detention time of five days. Batch operations should include holding times of at least 23 hours at 50°C, ten hours at 55°C or 4 hours at 60°C. When followed by anaerobic digestion, the aerobic thermophilic stage should meet the requirements for sludge pasteurization (section 2.2.4.2) or maintain 60°C for at least four hours with no addition of raw sludge, and the anaerobic stage must operate at a temperature not less than 30°C.

9.5.3 The FUCHS system

Using well-insulated reactors and a high-efficiency aeration system in aerobic thermophilic digestion of sewage sludge, the heat produced by the exothermic metabolic activity of the aerobic micro-organisms not only enables the temperature required for disinfection, at least 50°C, to be achieved but can also be sufficient to heat buildings housing the plant (Jakob *et al.* 1989). The sludge feed needs to contain at least 2.5% volatile solids, obtained by static thickening of the raw sludge. Depending on the type of sludge, 25 to 35% of organic material is removed in a minimum retention time of six days. The system operates as a two-stage process, to avoid the risk of any of the sludge short-circuiting the heat treatment and to ensure complete disinfection, but also confers operational stability. The FUCHS system described by Jakob *et al.* (1989) employs mechanical aerator–mixer and foam-control devices.

Approximate disinfection figures in the raw sludge, first stage (temperature range 35°C to 47°C) and second stage (47.5°C to maintained maximum 50°C), respectively (cited from an unpublished report by Strauch 1986) were:

- **Enterobacteriaciae**, raw sludge: 10^6 cfu/ml, 1st stage sludge: 5×10^4 cfu/ml, 2nd-stage sludge: 30 cfu/ml.
- **Faecal streptococci**, raw sludge: 10^5 cfu/ml, 1st-stage: 10^4 cfu/ml, 2nd-stage sludge 100 cfu/ml.
- **Enteroviruses**, raw sludge: 10^4 pfu/l, 1st-stage sludge: 3000 pfu/l, 2nd-stage sludge: 100 pfu/l.
- *Salmonellae* in 85% of samples of raw sludge, 1st-stage sludge samples: 35%, 2nd-stage sludge samples: none detected.
- *Ascaris* eggs found in about 5% of 1st-stage samples, none in 2nd-stage samples.

9.5.4 High-purity oxygen ATSD

Results on the use of high-purity oxygen in ATSD were reported by Booth & Tramontini (1984). Oxygen from an on-site pressure-swing adsorption (PSA) plant was used in investigating thermophilic aerobic digestion of sewage sludge in a 60-m^3 tank. Counts of 10^3 to 10^5/ml of total and faecal coliforms and faecal streptococci in the raw sludge were little affected by anaerobic digestion but completely eliminated by oxidative digestion at 68°C with a ten-day retention time, 69°C with a five-day retention time and even at 51°C with a five-day retention time. The *Salmonella* counts of about 20/ml in the raw sludge were completely removed with both five- and ten-day retention times at 68°C. It was concluded that temperatures of 65°C were self-sustaining in the system and that these were sufficient to eliminate virtually all pathogenic bacteria with the retention times used. Interestingly, it was found that the unreliability of dissolved-oxygen measurement made control of dissolved-oxygen levels difficult at 60°C.

In a process reported by Langeland *et al.* (1985), thickened sewage sludge was stabilized by aerobic thermophilic digestion in a full-scale three-stage process using high-purity oxygen. Different retention times and working temperature ranges were tested, and the treated sludge was examined for indicator organisms normally present in human and animal faeces, faecal coliforms, faecal streptococci and spores of *Clostridium perfringens*, bacteriophage coliphage MS2 and a pathogenic bacterium, *Salmonella* (principally *Salmonella oranienberg*).

With the working temperature in the range 60 to 65°C, faecal coliform and *Salmonella* were inactivated using mean retention times of five, seven and ten days, but faecal streptococci tended to survive the treatment, and *Clostridium perfringens* spores were little affected (reduction by about one log-decade). When temperatures around 50°C were used, all the test organisms survived treatment to a greater or lesser extent.

9.6 WET AIR OXIDATION

Disinfection, recontamination and regrowth of indicator bacteria in a wastewater sludge heat treatment system has been reported (Fujioka *et al.* 1988, Loh *et al.* 1988). Primary sludge, produced in a Hawaiian sewage-treatment works at a rate of 25 400 kg(dry solids)/day, is macerated and then heat-treated using the Zimpro process at 190°C for 30 minutes at a pressure by 22 atm (2230 kPa). Assays of total coliforms, faecal streptococcus, *Clostridium perfringens* and infectious human enteric viruses on sludge samples showed that the heat treatment generally disinfected the primary sludge, but the centrifuged sludge cake often contained high concentrations of faecal indicator organisms. This was shown to be due to contamination of the heat-treated sludge by raw sludge and by chlorinated primary effluent, and to regrowth of indicator bacteria in sludge storage tanks while awaiting centrifugation.

9.7 IRRADIATION

Doses giving 90% inactivation are called D_{10} values (White 1984). D_{10} values for bacteria are generally 5 to 30 krad, except that *Streptococcus faecalis* require 125 krad and *Clostridium perfringens* 200 krad. Values for spores are *Aspergillus niger* 31 krad and *Clostridium perfringens* 200 krad. The inactivation of faecal streptococci was found to be ten times greater in digested sludge than in raw sludge for the same dose of 0.4 Mrad.

In a plant at Geiselbullach, near Munich, Germany, irradiation of sewage sludge with ^{60}Co at a dose of 300 krad reduced total bacterial counts a hundredfold (2 log-decades), faecal streptococci also by a hundredfold (2 log-decades) and *Salmonellae* and other enterobacteria (in culture medium) by 5 to 6 log-decades (Havelaar 1984).

The effectiveness of irradiation is increased at higher temperatures. Re-contamination of irradiated sludge had not in 1984 been widely investigated, but irradiation does apparently improve the de-watering properties of the sludge.

9.8 LIME TREATMENT

Tests on lime stabilization of sewage sludge for disposal to agricultural land were reported by Beaumont (1984). Approximately 6 m^3 of a 10% w/v slurry of hydrated lime are added to 100 m^3 raw sludge to give a pH value above 12 and producing a slight odour of ammonia. The raw sludge contains 80 to 400 *Salmonella* organisms per litre of sludge, 4×10^8 *Escherichia coli*/l and 1.3×10^9 coliforms/l. Freshly mixed sludge and lime after two hours' contact time had no *Salmonella* and a coliform count of 2500/l; the counts in the lime-stabilized sludge for land application were typically no *Salmonella*, coliforms 1500/l and *E. coli* 1000/l.

Quinn and Hall (1984) reported tests on lime stabilization of undigested sewage sludge. Up to 98% of all bacteria are destroyed at pH 9.5 to 10.0 (7% lime on dry solids), increasing to 99.5% at pH 12. A persistent 0.5 to 1% remained even after very high lime doses, attributed to survival of endospores of spore-forming bacteria. Lime treatment greatly reduces virus viability, with less than 0.01% surviving pH 12. The temperature–time data given are not altogether clear, but appear to comprise five to 12 days at an ambient temperature of 15°C.

Problems with lime-treatment of sewage sludges were encountered in Norway as a result of difficulties in de-watering high-pH sludge and the recirculation of the high-pH liquors recovered (Paulsrud & Eikum 1984). Better results were obtained with addition of quicklime to de-watered sludge. Counts of coliforms and faecal streptococci were decreased by storage alone of sludge, without the addition of lime, and low lime dosage, 50 g($Ca(OH)_2$)/kg(sludge dry solids), had no additional effect. The counts were reduced to an undetectable level (<2 organisms/ml) by a dosage of 200 g(lime)/kg(sludge d.s.). It is also likely that viruses and thin-shelled parasite ova are inactivated at pH levels greater than 12, although thick-shelled ova are thought to survive lime treatment for long periods. The temperature rise resulting from the use of quicklime is probably critical for *Ascaris* ova.

In investigating the influence of lime treatment of raw sludge on the survival of pathogens, on the digestibility of sludge and on the production of methane (Pfuderer 1985), it was found that *Salmonella* organisms in raw sludge can be inactivated by lime treatment to give an initial pH of 12.8 and a contact time of three hours. Lime treatment was not found to inhibit subsequent anaerobic digestion of the sludge; although less biogas was generated, its methane content was increased.

The effect of lime on spores of *Bacillus anthracis* in the sludge of a treatment plant connected with some tanneries was reported by Lindner and Böhm (1985). All spores of *Bacillus anthracis* occurring in tannery effluent are likely to be concentrated in sludge, and the effect of lime treatment on their survival was investigated. Sludge was dosed with *Bacillus anthracis* spores, treated with slaked lime and quicklime at 10 and 20 kg/m^3 sludge, de-watered in a filter-press and stored for 23 weeks. A high proportion of spores survived both filtration and storage in all cases, although the actual pH levels achieved in the sludge cake were not reported.

Strauch (1989) concluded that lime treatment should produce a pH not less than 12.6. When slaked lime is used, lime treatment should be followed by storage for a minimum of three months before use. When quicklime is used, the temperature achieved, as a result of exothermic hydration of the lime, should reach at least 55°C for at least two hours.

9.8.1 Sewage treatment with lime

In physico-chemical treatment of wastewater, the use of lime-based coagulants in upflow blanket clarification has the potential to remove large proportions of bacteria, viruses and parasites (Smith 1989). Deactivation of bacteria is enhanced when treated effluent is diluted with seawater. The lime was applied in the form of a patented high-solids (50%) slurry, and was mixed with raw sewage in a coagulation vessel. The lime and sludge solids from a flocculent blanket which is removed in an upward-flow separator for de-watering and disposal.

In full-scale plant, operating at 20 000 m^3/day, and with a two-hour holding time following the lime treatment at pH 11.0, the average faecal coliform counts were reduced by a factor of over 2000 from 224 000/ml in the crude sewage. Over a 12-month sampling period, a 600-fold reduction in faecal coliforms was achieved using pH 10.5. Complete removal of *Salmonella* was achieved at both pH 9.5 and 10.5; virus counts in the settled, lime-treated effluent were about a hundred per litre. Although the virus count in the crude sewage was not determined, it was assumed to be several thousand per litre.

Laboratory tests at pH 11.4 also indicated 100% removal of *Ascaris* and 99% removal of total parasites, 100% removal of *Salmonella*, 98.6% removal of simian rotavirus SA-11 and reduction of faecal coliform by 5 log-decades. After 500 to 1000-fold dilution with seawater, the time required for a 90% die-off of both total and faecal coliforms in lime-treated effluent, using pH 10 to 10.7, was half that of the raw sewage.

9.9 COMPOSTING

Composting has the advantage of being a low-technology process which can produce a pasteurized sludge in the form of a valuable horticultural soil. The process needs to be well-managed in order to ensure adequate treatment. In the UK, the process has the timely advantages that straw can be used as a bulking agent in composting, and that there is currently concern about the depradation of peat bogs for providing horticultural compost. Composting thus provides a useful and profitable means of utilizing the 10 million tons of straw that will require disposal each year after stubble burning is prohibited in the UK in 1992.

For windrow composting, the initial water content of the sludge-bulking agent mixture should be between 40 to 60% (Bertoldi *et al.* 1983). This is to prevent blockage of air access channels by water with consequent creation of anaerobic zones, while maintaining a water content sufficient for the biological reactions. The mixture should achieve a temperature of at least 55°C for at least three weeks.

For reactor composting, the initial water content should be less than 70% and achieve a temperature of at least 55°C for at least ten days, as well as at least 65°C for at least 48 hours. The product should then be matured in windrows for two weeks.

Several different sludge composting systems were evaluated by Bertoldi *et al.* (1985) for pathogen inactivation. Different bulking agents, biodegradable solid municipal waste, rice bran, cork sawdust, wheat straw, wood chips and inert plastic spheres, were tested in open (windrow) composting of aerobically stabilized primary sewage sludge, in static heaps and turned heaps. The static heaps were subject to several different types of forced ventilation. Four closed reactors with forced ventilation, two horizontal and two vertical, were tested. The horizontal reactors were operated either statically or with the contents agitated, and the vertical reactors either continuously or discontinuously. Sludge samples were tested for *Salmonella* spp., faecal coliforms and faecal streptoccoci. Pathogen removal depends principally on the temperature achieved in the system. Static composting appears to be more effective than systems with turning, and in closed systems, the horizontal reactors appear to work better than the vertical reactors. This was attributed to better process control and homogeneity and more uniform oxygenation of the sludge mass.

Composting sludge usually requires the addition of a bulking agent to the sludge to correct the nutrient (carbon-to-nitrogen) ratio and to ensure adequate access of oxygen. Raw sewage sludge has, however, been successfully stabilized by aerobic composting without a bulking agent, in the form of filter-cake containing 20 to 22% dry matter (Coppola *et al.* 1989). The process was carried out in a closed, rotating tank supplied with high-purity oxygen, and reached disinfection temperatures, above 60°C, in less than 12 hours. Microbial species isolated from the compost varied with the temperature.

Composting at 45°C

73% of bacterial strains isolated were sporeformers, numbering 99 strains of sporeformers, 36 strains of non-sporeformers, seven of Actinomycetes and five of Eumycetes.

Bacillus licheniformis, B. coagulans (seven strains), *B. subtilis* (15 strains), *B. circulans* (nine strains), *B. brevis* (54 strains), *B. sphaericus* (13 strains). *Micropolyspora* sp. (two strains), *Thermomonospora* sp. (two strains), *Thermoactinomyces* sp. (two strains), *Streptomyces* sp., *Talaromyces emersonii* (two strains), *Aspergillus* sp., *Penicillium* sp., *Fusarium* sp.

Composting at 55°C

82% of bacterial strains isolated were sporeformers, numbering 63 strains of sporeformers, 13 of non-sporeformers, five of Actinomycetes, two of Eumycetes.

Bacillus licheniformis (23 strains), *B. stearothermophilus* (18 strains), *B. coagulans* (22 strains), *Micropolyspora* sp. (two strains), *Thermomonospora* sp., *Thermoactinomyces* sp. (two strains), *Talaromyces emersonii* (two strains).

Composting at 62°C

All bacteria isolated were sporeformers, numbering 83 strains.

Bacillus stearothermophilus (38 strains), and 45 strains of other *Bacillus* sp.

9.10 CONCLUSIONS

1. From the results with reported processes, it is clear that sludge needs to be subjected to adequate heat treatment in order to ensure destruction of all

pathogenic organisms, including helminth ova, protozoal cysts and bacterial spores. This may be achieved by thermophilic digestion, aerobic or anaerobic, or by including a pasteurization stage in conventional sludge treatments.

2. If current research demonstrates the prions could be significant infective agents in human or animal wastes, the severity of sludge treatments would need to be dramatically increased.
3. The processes need to be well-designed and well-operated so as to ensure that all elements of the sludge are subjected to the required treatment, avoiding hydraulic short circuiting and ensuring complete mixing, and that the risk of re-infection of treated sludge with raw sludge is obviated.
4. Anaerobic digestion has the advantage of generating a useful fuel gas, although in temperate climates, some or all of this may have to be used to maintain thermophilic temperatures. The generation of combustible gas inevitably involves an explosion hazard. Anaerobically treated sludge is stabilized and valuable as a fertilizer, but when applied to pasture, an appropriate period of time should be allowed after application of sludge before grazing of animals.
5. Thermophilic aerobic digestion can be operated with no additional heat input in well-designed and well-managed systems, apart from electrical energy for pumping and air compression. There can be odour problems with the exhaust gas however.
6. The risk of re-infection would be considerably reduced if whole sewage were subjected to aerobic secondary treatment, instead of the usual practice of mixing pathogen-depleted secondary sludge with pathogen-laden primary sludge. This would, however, require an approximate doubling of aeration capacity in a sewage treatment plant.
7. Composting can produce a valuable horticultural product from sludge, which could replace peat and alleviate depradation of peat bogs. It could also provide a means of utilizing the waste straw that will require disposal after stubble-burning is banned in the UK in 1992. The composting process needs very careful management, however, to ensure that sludge pathogens are destroyed, and on this basis appears to be rather less reliable than aerobic or anaerobic thermophilic digestion.
8. Entry into sewers and sewage of substances inhibitory to sludge treatment, or, indeed, sewage treatment generally, should be prevented or severely reduced, notably heavy metals, chlorinated hydrocarbons and anionic detergents, as well as persistent organic compounds, such as PCBs, that pass through treatments unchanged and remain in the final treated sludge.

References

Abel, P. D. (1989) *Water Pollution Biology*, Chichester, Ellis Horwood.

Adams, M. R. & Lloyd, D. A. (1989) Rotavirus as a viral indicator in shellfish. *WATERSHED '89: The Future for Water Quality in Europe*, eds Wheeler, D., Richardson, M. L. & Bridges, J., IAWPRC/Pergamon, 397–404.

Aiba, S. & Sudo, R. (1965) Parasites in sewage and possibilities of their extinction. *Adv. Water Polln. Res.*, **2**, 282–284.

Akin, E. W. & Jakubowski, W. (1986) Drinking water transmission of giardiasis in the United States. *Water Science & Technology*, **18**(10), 219–226.

Akin, E. W., Jakubowski, W., Lucas, J. B. & Pahren, H. R. (1977) Health hazards associated with wastewater effluents and sludge: microbiological considerations. Proceedings of conference *Risk Assessment and Health Effects of Land Application of Municipal Wastewater and Sludges*, San Antonio, Texas, USA, 12–14th December.

al-Ghazali, M. R. & al-Azawi, S. K. (1988a) Effects of sewage treatment on the removal of *Listeria monocytogenes*. *J. Appl. Bact.*, **65**, 203–208.

al-Ghazali, M. R. & al-Azawi, S. K. (1988b) Storage effects of sewage sludge cake on the survival of *Listeria monocytogenes*. *J. Appl. Bact.*, **65**, 209–213.

Anderson, D. R. & Garber, W. F. (eds) (1990) *IAWPRC Sludge Management Conference*, Los Angeles, January.

Anderson, R. M. & May, R. M. (1979) Population biology of infectious diseases: part I. *Nature*, **280**, 361.

Araujo, R. M., Arribas, R. M., Lucena, F. & Pares, R. (1989) Relation between *Aeromonas* and faecal coliforms in fresh waters. *J. Appl. Bact.*, **67**, 213–217.

Argent, V. A., Bell, J. C. & Edgar, D. (1981) Animal disease hazards of sewage-sludge disposal to land: effects of sludge treatment on *Salmonella*. *Water Pollution Control*, **80**(4), 537–540.

Arimi, S. M., Fricker, C. R. & Park, R. W. A. (1986) The occurrence of campylobacters in sewage and their removal by sewage treatment processes. *J. Appl. Bact.*, **61**, 18.

Arkhipova, N. S. (1979) Role of pasture irrigation with sewage effluent in the epizootiology of cattle cysticercosis. *Veterinary Bulletin*, **49**, 335 (cited by Snowdon *et al*. 1989a).

Arther, R. G., Fitzgerald, P. R. & Fox, J. C. (1981) Parasite ova in anaerobically digested sludge. *J. Water Pollution Control Federation*, **53**(8), 1334–1338.

Arundel, J. H. (1972) A review of cysticercosis of sheep and cattle in Australia. *Australian Vet. J.*, **48**, 140–155.

Arundel, J. H. & Adolph, A. J. (1960) Preliminary observations on the removal of *Taenia saginata* eggs from sewage using various treatment processes. *Australian Vet. J.*, **56**, 492–495.

Bailey, S. W. (1990) Management of livestock manure. In Cryptosporidium *in Water Supplies*, Department of the Environment & Department of Health, London, HMSO, 100–106.

Barcina, I., González, J. M., Irriberri, J. & Egea, L. (1990) Survival strategy of *Escherichia coli* and *Enterococcus faecalis* in illuminated fresh and marine systems. *J. Appl. Bact.*, **68**, 189–198.

Barnard, J. L., du Plessis, J. S. & Hosford, W. D. (1990) The active sludge pasteurisation process. *IAWPRC Sludge Management Conference*, Los Angeles, January, eds Anderson, D. R. & Garber, W. F.

Beaumont, F. (1984) An acceptable alternative: lime stabilisation of sewage sludge and disposal to agricultural land. *Sewage Sludge Stabilisation and Disinfection*, ed. Bruce, A. M., Chichester, Ellis Horwood, 529–540.

Berg, G. & Berman, D. (1980) Destruction by anaerobic and thermophilic digestion of indicator bacteria indigenous to domestic sludges. *Appl. Environm. Microbiol.*, **39**, 361–368.

Berg, G., Sullivan, G. & Venosa, A. D. (1988) Low-temperature stability of viruses in sludges. *Applied & Environmental Microbiol.*, **54**(3), 839–841.

Berman, D., Rice, E. W. & Hoff, J. C. (1988) Inactivation of particle-associated coliforms by chlorine and monochloramine. *Applied & Environmental Microbiol.*, **54**(2), 507–512.

Bertoldi, M. de, Coppola, S. & Spinosa, L. (1983) Health implications in sewage sludge composting. *Disinfection of Sewage Sludge: Technical, Economic and Microbiological Aspects*, eds Bruce, A. M., Havelaar, A. H. & L'Hermite, P., Dordrecht & London, Reidel, 165–178.

Bertoldi, M. de, Fransinetti, S., Bianchi, L. & Pera, A. (1985) Sludge hygienization with different compost systems. *Inactivation of Microorganisms in Sewage Sludge by Stabilisation Processes*, eds Strauch, D., Havelaar, A. H. & L'Hermite, P., London, Elsevier Applied Science Publishers, 64–76.

Bhaskaran, T. R., Sampathkumaran, M. A., Sur, T. C. & Radhakrishnan, I. (1956) Studies on the effect of sewage treatment processes on the survival of intestinal parasites. *Indian J. Medical Research*, **44**, 1.

Bingham, A. K. & Jarroll, E. L. (1979) *Giardia* sp.: physical factors of excystation *in vitro*, and excystation *vs.* eosin exclusion as determinants of viability. *Exp. Parasitology*, **47**, 284–291.

Bird, A. F. & McClure, M. A. (1976) The tylenchid (Nematoda) egg shell: structure, composition and permeability. *Parasitology*, **72**, 19–28.

Birkhead, G. & Vogt, R. L. (1989) Epidemilogic surveillance for endemic *Giardia lamblia* infection in Vermont. *Amer. J. Epidemiol.*, **129**(4), 762–768.

Bitton, G. (1980) *Introduction to Environmental Virology*, Chichester, Wiley.

Black, M., Scarpino, P. V., O'Donnell, C. J., Meyer, K. B., Jones, J. V. & Kaneshiro, E. S. (1982) Survival rates of parasite eggs in sludge during aerobic and anaerobic digestion. *Applied & Environmental Microbiology*, **44**(5), 1138–1143.

Black, R. E., de Romaña, G. L., Brown, K. H., Bravo, N., Bazador, O. G. & Kanashiro, H. C. (1989) Incidence and etiology of infantile diarrhoea and major routes of transmission in Huascar, Peru. *Amer. J. Epidemiol.*, **129**(4), 785–799.

Blamire, R. V., Goodhand, R. H. & Taylor, K. C. (1980) A review of some animal diseases encountered at meat inspections in England and Wales 1969 to 1978. *Veterinary Record*, **106**, 195–199.

Blaser, M. J., Smith, P. F., Wang, W.-L. L. & Hoff, J. C. (1986) Inactivation of *Campylobacter jejuni* by chlorine and monochloramine. *Appl. & Environmental Microbiology*, **51**(2), 307–311.

Bøckman, O. Chr. & Bryson, D. D. (1989) Well-water methaemoglobinaemia: the bacterial factor. *WATERSHED '89: The Future for Water Quality in Europe*, eds Wheeler, D., Richardson, M. L. & Bridges, J., IAWPRC/Pergamon, **2**, 239–244.

Bolton, F. J., Coates, D., Hutchinson, D. N. & Godfree, A. F. (1987) A study of thermophilic campylobacters in a river system *J. Appl. Bact.*, **62**, 167–176.

Booth, M. G. & Tramontini, E. (1984) Thermophilic sludge digestion using oxygen and air. Ch. 15 in *Sewage Sludge Stabilisation and Disinfection*, ed. Bruce, A. M., Chichester, Ellis Horwood, 293–311.

Bosch, A., Lucena, F. & Jofre, J. (1986) Fate of human enteric viruses (rotaviruses and enteroviruses) in sewage after primary sedimentation. *Water Science & Technology*, **18**(10), 47–52.

Bosch, A., Pinto, R. M., Blanch, A. R. & Jofre, J. T. (1988) Detection of human rotavirus in sewage through two concentration procedures. *Water Research*, **22**(3), 343–348.

Bradley, D. J. (1977) Health aspects of water supplies in tropical countries. *Water, Wastes and Health in Hot Climates*, eds Feachem, R., McGarry, M. & Mara, D., Wiley.

Brandt, I. R. A. & Sewell, M. M. H. (1981) *In vitro* hatching and activation of *Taenia taeniaformis* oncospheres. *Veterinary Res. Comm.*, **5**, 193–199.

Brown, H. W. (1928) A quantitative study of the influence of oxygen and temperature on the embryonic development of the eggs of the pig ascarid *Ascaris suum* Goetze, *J. Parasitology*, **14**(3), 141–160.

Brown, P. (1990) Britain to clean up its act for Europe. *The Guardian*, Tuesday 6th March, 20.

Browning, J. R. & Ives, D. G. (1987) Environmental health and the water distribution system: a case history of an outbreak of giardiasis, *J. Inst. Water & Environmental Management*, **1**(1), 55–60.

Bruce, A. M. (ed.) (1984) *Sewage Sludge Stabilisation and Disinfection*, Chichester, Ellis Horwood.

Bruce, A. M. (1989) Other investigations of thermophilic aerobic digestion in the UK. *Treatment of Sewage Sludge*, eds Bruce, A. M., Colin, F. & Newman, P. J., London, Elsevier Applied Science, 39–43.

Bruce, A. M., Colin, F. & Newman, P. J. (eds) (1989) *Treatment of Sewage Sludge*, London, Elsevier Applied Science.

Bruce, A. M., Pike, E. B., Fisher, W. J. (1990) A review of treatment process options to meet the EC sludge directive. *J. Inst. Water & Environmental Management*, **4**, 1–13.

Brunner, P. H. & Lichtensteiger, Th. (1989): Landfilling of sewage sludge — practice and legislation in Europe, *Treatment of Sewage Sludge*, eds Bruce, A. M., Colin, F. & Newman, P. J., London, Elsevier Applied Science, 52–57.

Buras, N. (1976) Concentration of enteric viruses in wastewater and effluent: a two-year survey. *Water Research*, **10**, 295–298.

Burge, W. D., Cramer, W. N. & Epstein, E. (1978) Destruction of pathogens in sewage sluge by composting. *Trans ASEA*, **21**, 510–514.

Bürger, H.-J. (1983) Survival of *Taenia* eggs in sewage and on pasture. Proc. 3rd International Symposium *Processing and Use of Sewage Sludge*, Brighton, UK, 27–30th September.

Butler, R. C., Lund, V. & Carlson, D. A. (1987) Susceptibility of *Campylobacter jejuni* and *Yersinia enterocolitica* to UV radiation. *Appl. & Environm. Microbiol.*, **53**(2), 375–378.

Byram, J. E. & Senft, A. W. (1979): Structure of the schistosome egg shell: amino-acid analysis and incorporation of labelled amino-acids. *American J. Tropical Med. & Hygiene*, **28**(3), 539–547.

Campbell, H. W. (1989) A status report on Environment Canada's oil from sludge technology. In *Sewage Sludge Treatment and Use*, eds Dirkzwager, A. H. & L'Hermite, P., London, Elsevier Applied Science, 281–290.

Carrington, E. G. (1978) *The Contribution of Sewage Sludges to the Dissemination of Pathogenic Micro-organisms in the Environment*, Water Research Centre Technical Report TR71.

Carrington, E. G. (1980) *The Fate of Pathogenic Micro-organisms during Wastewater Treatment and Disposal*, Water Research Centre Technical Report TR128.

Carrington, E. G. (1985) Pasteurisation: effects upon *Ascaris* eggs. In *Inactivation of Microorganisms in Sewage Sludge by Stabilisation Processes*, eds Strauch, D., Havelaar, A. H. & L'Hermite, P., London, Elsevier Applied Science Publishers, 121–125.

Carrington, E. G. & Harman, S. A. (1984a) The effect of anaerobic digestion temperature and retention period on the survival of *Salmonella* and *Ascaris* ova. Ch. 19 in *Sewage Sludge Stabilisation and Disinfection*, ed. Bruce, A. M., Chichester, Ellis Horwood, 369–380.

Carrington, E. G. & Harman, S. A. (1984b) The effect of gamma radiation and subsequent storage upon *Salmonella* and other bacteria in sewage sludge. *Sewage Sludge Stabilization and Disinfection*, ed. Bruce, A. M., Chichester, Ellis Horwood, 546–549.

Carter, A. M., Pacha, R. E., Clark, G. W. & Williams, E. A. (1987) Seasonal occurrence of *Campylobacter* spp. in surface waters and their correlation with standard indicator bacteria. *Appl. & Environm. Microbiol.*, **53**(3), 523–526.

Casemore, D. P. (1989) *Cryptosporidium*: an emerging pathogen in the water industry. Symposium on *Water Pollution*: *Microbiology*, *Chemistry and Treatment*, Birmingham University, 27–28 September.

Casemore, D. P. (1990) Epidemiological aspects of human cryptosporidiosis. Cryptosporidium *in Water Supplies*, Department of the Environment and

Department of Health, 108–132, reproduced from *Epidemiology & Infection*, **104**, 1–28.

Chen, S., Redwood, D. W. & Ellis, B. (1990) Control of *Campylobacter fetus* in artificially contaminated bovine semen by incubation with antibiotics before freezing. *Brit. Vet. J.*, **146**, 68–74.

Chesshire, M. J. (1986) A comparison of the design and operational requirements for the anaerobic digestion of animal slurries and of sewage sludge. In *Anaerobic Digestion of Sewage Sludge and Organic Agricultural Wastes*, eds Bruce, A. M., Kouzeli-Katsiri, A. & Newman, P. J., London, Elsevier Applied Science Publishers, 33–54.

Clarke, A. J. & Perry, R. N. (1980) Egg shell permeability and hatching of *Ascaris suum*. *Parasitology*, **80**, 447–456.

Coppola, S., Villani, F. & Romano, F. (1989) Composting raw sewage sludge in the absence of bulking agents. *Sewage Sludge Treatment and Use*, eds Dirkzwager, A. H. & L'Hermite, P., London, Elsevier Applied Science, 433–439.

Council of the European Communities (1986) Council Directive on the protection of the environment, and in particular the soil, when sewage sludge is used in agriculture, 86/278/EEC, 12 June 1986. *Official Journal of the European Communities*, L181, **29** (4 July), 6–12.

Cram, E. B. (1926) Ascariasis in preventitive medicine. *American J. Tropical Medicine*, **6**, 91–94.

Cram, E. B. (1943) The effect of various treatment processes on the survival of helminth and protozoan cysts in sewage. *Sewage Works J.*, **15**, 1119–1138.

Crewe, W. (1977) Human health, Department of the Environment research seminar on pathogens in sewage sludge. *Water Engineering* **1**, Research & Development Division, Technical Note No. 7.

Crewe, W. (1983) The transmission of *Taenia saginata*. Symposium on *Trends in Research on Cestode Infections*, Liverpool School of Tropical Medicine, 22–23 November.

Crewe, W. & Haddock, D. R. W. (1985) *Parasites and Human Disease*, London, Edward Arnold.

Crewe, W. & Owen, R. R. (1978) 750 000 eggs a day — £750 000 a year. *New Scientist*, **80**(1127), 344–346.

Crewe, W. & Owen, R. R. (1979) The transmission of bovine cysticercosis in Great Britain. *Trans. Royal Soc. Tropical Med. & Hygiene*, **73**, 344–346.

Cunningham, J. D. & Skinwood, J. F. (1990) Microbiological and toxics investigations on irradiated and unirradiated sewage sludge. *IAWPRC Sludge Management Conference*, Los Angeles, January, eds Anderson, D. R. & Garber, W. F.

Dagan, R., Zaltzstein, E. & Gorodischer, R. (1988) Methaemoglobinaemia in young infants with diarrhoea. *Eur. J. Paediatr.*, **147**, 87–89.

Dahling, D. R., Safferman, R. S. & Wright, B. A. (1989) Isolation of enterovirus and reovirus from sewage and treated effluents in selected Puerto Rican communities. *Appl. & Environm. Microbiol.*, **55**(2), 503–506.

Davis, R. D. (1987) Use of sewage sludge on land in the United Kingdom. In *Use of Soil for Treatment and Final Disposal of Effluents and Sludge*, eds Oliveira, P. R. C. & Almeida, S. A. S., *Water Science & Technology*, **19**(8), 1–8.

Davis, R. D. (1989) Agricultural utilisation of sewage sludge: a review. *J. Instn. Water & Environm. Management*, **3**, 351–355.

Deming, M. S., Tauxe, R. V., Blake, P. A., Dixon, S. E., Fowler, B. S., Jones, T. S., Lockamy, E. A., Patton, C. M. & Sikes, R. O. (1987) *Campylobacter* enteritis at a university: transmission from eating chicken and from cats. *Amer. J. Epidemiol.* **126**(3), 526–534.

Department of the Environment (1989) *Code of Practice for Agricultural Use of Sewage Sludge*, London, HMSO.

Department of the Environment & Department of Health (1990) Cryptosporidium *in Water Supplies*, London, HMSO.

Dirkzwager, A. H. & L'Hermite, P. (eds) (1989) *Sewage Sludge Treatment and Use*, London, Elsevier Applied Science.

Donnelly, F. A., Appleton, C. C. & Schutte, C. H. J. (1984): The influence of salinity on the ova and miracidia of three species of *Schistosoma*. *Int. J. Parasitology*, **14**(2), 113–120.

Dudley, D. J., Güntzel, M. N., Ibarra, M. J., Moore, B. E. & Sagik, B. P. (1980) Enumeration of potentially pathogenic bacteria from sewage sludges. *Appl. & Environm. Microbiol.*, **39**, 118–126.

Englund, P. T. & Sher, A. (eds) (1988) *The Biology of Parasitism*, New York, Alan R. Liss, Inc.

Enigk, K., Holl, P. & Dey-Hazra, A. (1975) The destruction of parasite resistant stages in sludge by irradiation with low accelerating voltage electrons. *Zbl. Bakt. Hyg.*, I. Abt. Orig. B, **161**, 61–71.

Evison, L. M. (1988) Comparative studies on the survival of indicator organisms and pathogens in fresh and sea water. *Water Science & Technology*, **20**(11/12), 309–315.

Evison, L. M. & James, A. (1977) Microbiological criteria for tropical water quality. *Water, Wastes and Health in Hot Climates*, eds Feachem, R. G., McGarry, M. & Mara, D., Wiley.

Fairbairn, D. (1961) The *in vitro* hatching of *Ascaris lumbricoides* eggs. *Can. J. Zoology*, **39**, 153–162.

Farrell, J. B. (1984) Recent developments in sludge digestion in the United States and a view of the future. Ch. 16 in *Sewage Sludge Stabilisation and Disinfection*, ed. Bruce, A. M., Chichester, Ellis Horwood, 317–329.

Farrell, J. B., Erlap, A. E., Rickabaugh, J., Freedman, D. & Hayes, S. (1988) Influence of feeding procedure on microbial reductions and performance of anaerobic digestion. *J. Water Polln. Control Fedn.*, **60**(5), 635–644.

Fayer, R. & Ungar, B. L. P. (1986) *Cryptosporidium* and cryptosporidiosis. *Microbiol. Rev.*, **50**, 458–483.

Feachem, R. G., Bradley, D. G., Garelick, H. & Mara, D. (1980) Appropriate technology for water supply and sanitation **3**. *Health Aspects of Excreta and Sullage Management*, Washington, D.C., USA, The World Bank.

Feachem, R. G., Bradley, D. G., Garelick, H. & Mara, D. (1983) *Sanitation and Disease*, Chichester, Wiley.

Fiechter, A. & Sonnleitner, B. (1989) Thermophilic aerobic stabilisation. *Sewage Sludge Treatment and Use*, eds Dirkzwager, A. H. & L'Hermite, P., London, Elsevier Applied Science, 291–301.

Fitzgerald, P. R. (1982) Helminth transmission from anaerobically digested sludge. *Report of the 5th International Conference on Parasitology*, 7–14 August, 290.

Fitzgerald, P. R. & Ashley, R. F. (1977) Differential survival of *Ascaris* ova in wastewater sludge. *J. Water Polln. Control Fed.*, **49**, 1722–1724.

Fitzgerald, P. R. & Prakasam, T. B. S. (1978) Survival of *Trichinella spiralis* larvae in sewage sludge anaerobic digesters. *J. Parasitology*, **64**, 445–447.

Flehmig, B., Billing, A., Vallbracht, A. & Botzenhart, K. (1985) Inactivation of hepatitis A virus by heat and formaldehyde. *Water Science & Technology*, **17**(10), 43–45.

Förstner, M. J. (1968) Investigations on the viability of worm eggs in sludge from digestion towers used in agriculture. *Zeitschrift Wasser u. Abwasser Forschung*, **2**(2), 57–61.

Förstner, M. J. (1970a) The effect of sewage overlying liquor and composting on the viability of parasitic reproductive stages. *Zeitschrift Wasser u. Abwasser Forschung*, **3**, 176–184.

Förstner, M. J. (1970b) Investigations on the effect of sludge digestion in Emscher tanks in main sewage works on the viability of worm eggs. *Zeitschrift Wasser u. Abwasser Forschung*, **3**, 57–58.

Fox, J. C. & Fitzgerald, P. R. (1976) Parasitic organisms present in sewage systems of a large sewage district. *J. Parasitology* (suppl.), **62**, 28.

Fricker, C. R. & Park, R. W. A. (1989) A two-year study of the distribution of 'thermophilic' campylobacters in human, environmental and food samples from the Reading area with particular reference to toxin production and heat-stable, serotype. *J. Appl. Bact.*, **66**, 477–490.

Fujioka, R. S., Hirano, W. M. & Loh, P. C. (1988) Disinfection, recontamination and regrowth of indicator bacteria in a wastewater sludge heat treatment system. *Water Science & Technology*, **20**(11/12), 329–335.

Galbraith, N. S., Barrett, N. J. & Stanwell-Smith, R. (1987) Water and disease after Croydon: a review of water-borne and water-associated disease in the UK 1937–86. *J. Instn. Water & Environmental Management*, **1**(1), 7–21.

Gallie, G. J. & Sewell, M. M. H. (1970) A technique for hatching *Taenia saginata* eggs. *Veterinary Record*, **86**, 749.

Gealt, M. A. (1988) Recombinant DNA plasmid transmission to indigenous organisms during waste treatment. *Water Science & Technology*, **20**(11/12), 179–184.

Gemmell, M. A. & Johnstone, P. D. (1977) Experimental epidemiology of hydatidosis and cysticercosis. *Adv. Parasitology*, **15**, 311–369.

Gerba, C. P. & Goyal, S. M. (1988) Enteric virus: risk assessment of ocean disposal of sewage sludge. *Water Science & Technology*, **20**(11/12), 25–31.

Godfree, A. F., Jones, F., Satchwell, M. & Watson, D. C. (1984) The effectiveness of chemical disinfection on faecal bacteria in sludge. Ch. 22 in *Sewage Sludge Stabilisation and Disinfection*, ed. Bruce, A. M., Chichester, Ellis Horwood, 412–425.

Gotaas, H. B. (1953) Reclamation of municipal refuse by composting, University of California, Sanitary Engineering Research Laboratory, Technical Bulletin No. 9, Series 37.

Grabow, W. O. K. (1968) The Virology of Wastewater Treatment. *Water Research*, **2**, 675–701.

Grabow, W. O. K. (1986) Indicator systems for assessment of the virological safety of treated drinking water, *Water Science & Technology*, **18**(10), 159–165.

Grabow, W. O. K., Coubrough, P., Hilner, C. & Bateman, B. W. (1984) Inactivation of hepatitis A virus, other enteric viruses and indicator organisms in water by chlorine. *Water Science & Technology*, **17**, 657–664.

Graham, A. F. & Lund, B. F. (1986) The effect of citric acid on growth of proteolytic strains of *Clostridium botulinum*. *J. Appl. Bact.*, **61**, 18.

Graham, H. J. (1981) Parasites and the land application of sewage sludge. *Canada-Ontario Agreement*, Research Report No. 110, Ottawa.

Granum, P. E. (1990) *Clostridium perfringens* toxins involved in food poisoning. *Int. J. Food Microbiol.*, **10**, 101–112.

Gregory, M. W. (1990) Epidemiology of cryptosporidiosis in animals. Cryptosporidium *in Water Suplies*, Department of the Environment & Department of Health, London, HMSO, 82–99.

Grüttner, H. & Larsen, A. B. (1990) The use of fermentation sludge as fertiliser. *IAWPRC Sludge Management Conference*, Los Angeles, January, eds Anderson, D. R. & Garber, W. F., Friday Session 4, Paper 2.

Gyorkos, T. W., Frappier-Davignon, L., MacLean, J. D. & Viens, P. (1989) Effect of screening and treatment on imported intestinal parasite infections: results from a randomized controlled trial. *Amer. J. Epidemiol.*, **129**(4), 753–761.

Hall, J. E. (1989a) Methods of applying sewage sludge to land. A review of recent developments. In *Sewage Sludge Treatment and Use*, eds Dirkzwager, A. H. & L'Hermite, P., London, Elsevier Applied Science, 65–84.

Hall, J. E. (1989b) Treatment and environmental impact of disposal of sewage sludge. Symposium on *Water Pollution*: *Microbiology*, *Chemistry and Treatment*, Birmingham University, 27–28 September.

Hamer, G. (1989) Fundamental aspects of aerobic thermophilic biodegradation. In *Treatment of Sewage Sludge*, eds Bruce, A. M., Colin, F. & Newman, P. J., London, Elsevier Applied Science, 2–19.

Hamlin, E. J. (1946) Sewage disposal as a national problem. *The Surveyor*, **105**, 919.

Hancock, D. D., Wikse, S. E., Lichtenwalner, A. B., Westcott, R. B. & Gay, C. C. (1989) Distribution of bovine cysticercosis in Washington. *Amer. J. Vet. Res.*, **50**(4), 564–570.

Harf, C. & Monteil, H. (1988) Interactions between free-living amoebae and *Legionella* in the environment. *Water Science & Technology*, **20**(11/12), 235–239.

Hartmann, G., Eller, J. & Gloyna, E. F. (1989) Water oxidation of sludges and toxic wastes. In *Environmental Engineering*, Proc. 1989 Specialty Conference, ed. Malina, J. F., Amer. Soc. Civil Engineers, 804–810.

Havelaar, A. H. (1984) Sludge disinfection — an overview of methods and their effectiveness, Ch. 2 in *Sewage Sludge Stabilization and Disinfection*, ed. Bruce, A. M., Chichester, Ellis Horwood, 48–60.

Havelaar, A. H. (1985) Conclusions. From *Inactivation of Microorganisms in Sewage Sludge by Stabilisation Processes*, eds Strauch, D., Havelaar, A. H. & L'Hermite, P., London, Elsevier Applied Science Publishers, 189–190.

Havelaar, A. H. & Block, J. C. (1986) Epidemiological studies related to the use of sewage sludge in agriculture. *Processing and Use of Organic Sludge and Liquid Agricultural Wastes*, ed. L'Hermite, P., Dordrecht, Reidel, 210–214.

Havelaar, A. H. & Pot-Hogeboom, W. M. (1988) F-specific RNA-bacteriophages as model viruses in water hygiene: ecological aspects. *Water Science & Technology*, **20**(11/12), 399–407.

Havelaar, A. H. & van Olphen, M. (1989) Water quality standards for bacteriophages? *WATERSHED '89: The Future for Water Quality in Europe*, eds Wheeler, D., Richardson, M. L. & Bridges, J., IAWPRC/Pergamon, **2**, 357–366.

Herson, D. S., McGonigle, B., Payer, M. A. & Baker, K. H. (1987) Attachment as a factor in the protection of *Enterobacter cloacae* from chlorination. *Appl. & Environm. Microbiol.*, **53**(5), 1178–1180.

Höller, C. (1988) Long-term study of occurrence, distribution and reduction of *Campylobacter* spp. in the sewage system and wastewater treatment plant of a big town. *Water Science & Technology*, **20**(11/12), 529–531.

Huber, J. & Mihalyfy, E. (1984) Experiences with the pre-pasteurisation of sewage sludge with heat recovery. Ch. 20 in *Sewage Sludge Stabilisation and Disinfection*, ed. Bruce, A. M., Chichester, Ellis Horwood, 381–398.

Hudson, J. A. & Fennel, H. (1980) Disposal of sewage sludge to land: chemical and microbiological aspects of sludge to land policy. *Water Pollution Control*, **79**(3), 370–387.

Hughes, D. L., Morris, D. L., Norrington, I. J. & Waite, W. M. (1985) The effects of pasteurisation and stabilisation of sludge on *Taenia saginata* eggs. In *Inactivation of Microorganisms in Sewage Sludge by Stabilisation Processes*, eds Strauch, D., Havelaar, A. H. & L.'Hermite, P., London, Elsevier Applied Science Publishers, 126–134.

Hurst, C. J. (1988) Effect of environmental variables on enteric virus survival in surface freshwaters. *Water Science & Technology*, **20**(11/12), 473–476.

Jakob, J., Roos, H.-J. & Siekmann, K. (1989) Aerobic-thermophilic methods for disinfecting and stabilising sewage sludge. *Sewage Sludge Treatment and Use*, eds Dirkzwager, A. H. & L'Hermite, P., London, Elsevier Applied Science, 378–399.

Jakubowski, W. (1984) Detection of *Giardia* cysts in drinking water — the state of the art. In *Giardia and Giardiasis — Biology, Pathogenesis and Epidemiology*, eds Erlandsen, S. L. & Meyer, E. A., New York, Plenum Press, 263–286.

Jarroll, E. L., Bingham, E. K. & Meyer, E. A. (1981) Effect of chlorine on *Giardia lamblia* cyst viability. *Appl. & Environm. Microbiol.*, **41**, 483–487.

Jeffrey, H. C. & Leach, R. M. (1975, 1978) *Atlas of Medical Helminthology and Protozoology*, Edinburgh, Churchill & Livingstone, 1st and 2nd editions.

Jenkins, D. & Olson, B. H. (eds) (1989) *Water and Wastewater Microbiology*, IAWPRC/Pergamon.

Jepsen, A. & Roth, A. (1950) Parasitological probklem of sewage disposal. Occurrence of eggs of *Taenia saginata* in city sewage and distribution of *Cysticercus bovis* in calves. *Nor. Vet. Med.*, **2**, 967–991.

Jørgensen, P. H. & Lund, E. (1986) Transport of viruses from sludge application sites. *Processing and Use of Organic Sludge and Liquid Agricultural Wastes*, ed. L'Hermite, P., Dordrecht, Reidel, 215–224.

Kabler, P. W., Chang, S., Clarke, N. A. & Clarke, A. F. (1963) Pathogenic bacteria and viruses in water. *Proceedings 5th Sanitary Engineering Conference*, University of Illinois, Urbana.

Kabrick, R. M. & Jewell, W. J. (1982) Fate of pathogens in thermophilic aerobic sludge digestion. *Water Res.*, **16**, 1051–1060.

Keevil, C. W., West, A. A., Walker, J. T., Lee, J. V., Dennis, P. J. L. & Colbourne, J. S. (1989) Biofilms: detection, implications and solutions. *WATERSHED '89: The Future for Water Quality in Europe*, eds. Wheeler, D., Richardson, M. L. & Bridges, J., IAWPRC/Pergamon, **2**, 367–374.

Kelly, H. G., Mavinic, D. S., Koch, F. A., Wetter, R. D. & Melcer, H. (1990) Liquid composting of municipal sludge for agricultural use in small communities: Canadian application. *IAWPRC Sludge Management Conference*, Los Angeles, January, eds Anderson, D. R. & Garber, W. F., Friday, Session 4, Paper 1.

Ketchum, P. A. (1988) *Microbiology Concepts and Applications*, Wiley.

Kidson, R. J. & Ray, D. L. (1984) Pasteurisation by submerged combustion together with anaerobic digestion. Ch. 21 in *Sewage Sludge Stabilisation and Disinfection*, ed. Bruce, A. M., Chichester, Ellis Horwood, 399–408.

Kiff, R. J. & Lewis-Jones, R. (1984) Factors that govern the survival of selected parasites in sewage sludges. Ch. 25 in *Sewage Sludge Stabilisation and Disinfection*, ed. Bruce, A. M., Chichester, Ellis Horwood, 453–461.

Kiff, R. J., Cheung, Y. H., Brown, S. & Lewis-Jones, R. (1984) Sewage sludge detoxification in relation to sludge disposal strategy. *Sewage Sludge Stabilisation and Disinfection*, ed. Bruce, A. M., Chichester, Ellis Horwood, 576–580.

Knowlton, D. R. & Ward, R. L. (1987) Characterization of virucidal agents in activated sludge. *Appl. & Environm. Microbiol.*, **53**(4), 621–626.

Kong, L. I., Swango, L. J., Blagburn, B. L., Hendrix, C. M., Williams, D. E. & Worley, S. D. (1988) Inactivation of *Giardia lamblia* and *Giardia canis* cysts by combined and free chlorine. *Appl. & Environm. Microbiol.*, **54**(10), 2580–2582.

Krige, P. R. (1964) A survey of the pathogenic organisms and helminthic ova in compost and sewage sludge. *J. Inst. Sewage Purification*, **3**, 215–220.

Kutz, S. M. & Gerba, C. P. (1988) Comparison of virus survival in freshwater sources. *Water Science & Technology*, **20**(11/12), 467–471.

Langeland, G., Paulsrud, B. & Haugan, B.-E. (1985) Aerobic thermophilic stabilisation. *Inactivation of Microorganisms in Sewage Sludge by Stabilisation Processes*, eds Strauch, D., Havelaar, A. H. & L'Hermite, P., London, Elsevier Applied Science Publishers, 38–47.

Leahy, J. G., Rubin, A. J. & Sproul, O. J. (1987) Inactivation of *Giardia muris* cysts by free chlorine. *Appl. & Environm. Microbiol.*, **53**(7), 1448–1453.

LeChevallier, M. W., Cawthon, C. D. & Lee, R. G. (1988a) Mechanisms of bacterial survival in chlorinated drinking water. *Water Science & Technology*, **20**(11/12), 145–151.

LeChevallier, M. W., Cawthon, C. D. & Lee, R. G. (1988b) Factors promoting survival of bacteria in chlorinated water supplies. *Appl. & Environm. Microbiol.*, **54**(3), 649–654.

Lee, M.-R. & Shih, J. C. H. (1988) Effect of anaerobic digestion on oocysts of the protozoon *Eimeria tenella*. *Appl. & Environm. Microbiol.*, **54**(10), 2335–2341.

Leuze, M., Koch, K. & Wekerle, J. (1985) Influence of mesophilic anaerobic digestion and pasteurisation of raw and digested sludge on viruses occurring in humans and domestic animals. In *Inactivation of Microorganisms in Sewage Sludge by Stabilisation Proceses*, eds Strauch, D., Havelaar, A. H. & L'Hermite, P., London, Elsevier Applied Science Publishers, 2–27.

Levaillant, C. & Gallien, C. L. (1979) Sanitation methods using high energy electron beams. *Radiation Phys. Chem.*, **14**, 309–316 (cited by Waite *et al.*, 1989).

Lewis-Jones, R. & Kiff, R. J. (1984) A growing awareness of sewage pathogens. *Water & Waste Treatment J.*, **27**(2), 45.

Lindner, F. & Böhm, R. (1985) Effect of lime on spores of *Bacillus anthracis* in the sludge of a treatment plant connected with some tanneries. In *Inactivation of Microorganisms in Sewage Sludge by Stabilisation Processes*, eds Strauch, D., Havelaar, A. H. & L'Hermite, P., London, Elsevier Applied Science Publishers, 113–120.

Linfield, R. (1977) Potato cyst eelworm studies in the Anglian Water Authority. Research seminar on *Pathogens in Sewage Sludge*, Department of the Environment, Water Engineering Research & Development Division Technical Note 7.

Loh, P. C., Fujioka, R. S. & Hirano, W. M. (1988) Thermal inactivation of human enteric viruses in sewage sludge and virus detection by nitrose cellulose-enzyme immunoassay. *Chemical and Biological Characterisation of Sludges, Sediments, Dredge Spoils and Drilling Muds*, eds Lichtenberg, J. J., Winter, J. A., Weber, C. I. & Fradkin, L., ASTM *STP 976*, 273–281.

Loll, U. (1986) Biogas plants for animal slurries in the Federal Republic of Germany. In *Anaerobic Digestion of Sewage Sludge and Organic Agricultural Wastes*, eds Bruce, A. M., Kouzeli-Katsiri, A. & Newman, P. J., London, Elsevier Applied Science Publishers, 14–32.

Loll, U. (1989) Combined aerobic, thermophilic and anaerobic digestion of sewage sludge, *Treatment of Sewage Sludge*, eds Brucc, A. M., Colin, F. & Newman, P. J., London, Elsevier Applied Science, 20–28.

Lowe, H. N. Lacey, W. J., Surkiewicz, B. F. & Jaeger, R. F. (1956) Destruction of micro-organisms in water, sewage and sewage sludge by ionizing radiation. *J. Amer. Water Works Assn.*, **48**, 1362–1372.

Lund, E. (1975) Public hcalth aspects of waste-water treatment, Proceedings of the international symposium *The Use of High-level Radiation in Waste Treatment — Status and Prospects*, International Atomic Energy Agency, Munich, 17–21 March.

McAnally, A. S. & Benefield, L. D. (1989) Use of water hyacinths to upgrade treatment plants. In *Environmental Engineering*, ed. Malina, J. F., New York, ASCE, 208–213.

McCormack, D. M. & Spaull, A. M. (1987) *Studies of Potato Cyst Nematodes,* Globodera *spp. in Sewage Sludges*, Medmenham, Water Research Centre, Report PRS 16167-M.

MacDonald, R. J. & Ernst, A. (1987) Disinfection efficiency and problems associated with maturation ponds. *Water Science & Technology*, **18**(10), 19–29.

MacPherson, R. G., Mitchell, B. B. & McCance, C. B. (1978) Bovine cysticercosis storm following the application of human slurry. *Veterinary Record*, **102**, 156–157.

MacPherson, R. G., Mitchell, B. B. & McCance, C. B. (1979) Bovine cysticercosis storm following use of human slurry. *Meat Hygienist*, 21–32 (cited by Schwartzbrod *et al.*, 1987).

Mancini, P., Fertels, S., Nave, D. C. & Gealt, M. A. (1987) Mobilization of plasmid pHSV106 from *Escherichia coli* HB101 in a laboratory-scale waste treatment facility. *Appl. & Environm. Microbiol.*, **53**, 665–671.

Mara, D. (1978) *Sewage Treatment in Hot Climates*, Chichester, Wiley.

Martínez-Palomo, A. (1988) Biology of amebiasis: progress and perspectives. *The Biology of Parasitism*, eds Englund, P. T. & Sher, A., New York, Alan R. Liss, Inc., 61–76.

Mason, C. A. & Hamer, G. (1987) Survival and activity of *Klebsiella pneumoniae* at super-optimal temperatures. *Bioprocess Engineering*, **2**, 121–127 (cited in Hammer 1989).

Medhat, A. E. F. & Stafford, D. A. (1989) The fate of *Entamoeba histolytica* during sewage treatment using anaerobic digesters. *WATERSHED '89: The Future for Water Quality in Europe*, eds Wheeler, D., Richardson, M. L. & Bridges, J., IAWPRC/Pergamon, **2**, 381–390.

Merrett, H., Pattinson, C., Stackhouse, C. & Cameron, S. (1989) The incidence of enteroviruses around the Welsh coast — a three year intensive survey. *WATERSHED '89: The Future for Water Quality in Europe*, eds Wheeler, D., Richardson, M. L. & Bridges, J., IAWPRC/Pergamon, **2**, 345–351.

Mesquita, M. M. F. (1988) Effects of seawater contamination level and exposure period on the bacterial and viral accumulation and elimination processes by *Mytilus edulis*. *Water Science & Technology*, **20**(11/12), 265–270.

Milne, D. P. (1989) Shore-based microbiological sampling of recreational/bathing waters — possible problems and solution. *WATERSHED '89: The Future for Water Quality in Europe*, eds Wheeler, D., Richardson, M. L. & Bridges, J., IAWPRC/Pergamon, **2**, 331–336.

Mitsdörffer, R., Demharter, W. & Bischofberger, W. (1990) Stabilisation and disinfection of sewage sludge by two-stage anaerobic thermophilic-mesophilic digestion. *Water Science & Technology*, **22**(7/8), 289–290.

Morgan, S. F. & Gunson, H. G. (1989) The development of an aerobic thermophilic sludge digestion system in the UK. *Treatment of Sewage Sludge*, eds Bruce, A. M., Colin, F. & Newman, P. J., London, Elsevier Applied Science, 29–38.

Morgan, S. F., Gunson, H. G., Littlewood, M. H. & Winstanley, R. (1984) Aerobic thermophilic digestion of sludge using air. Ch. 14 in *Sewage Sludge Stabilisation and Disinfection*, ed. Bruce, A. M., Chichester, Ellis Horwood, 278–292.

Morris, R. (1989) Viruses and drinking water: a European perspective. Symposium on *Water Pollution: Microbiology, Chemistry and Treatment*, Birmingham University, 27–28 September.

Morris, R. & Cox, I. (1989) Virological quality of bathing waters in England. *WATERSHED '89: The Future for Water Quality in Europe*, eds Wheeler, D., Richardson, M. L. & Bridges, J., IAWPRC/Pergamon, **2**, 337–343.

Murray, K. C., Tong, A. & Bruce, A. M. (1990) Thermophilic anaerobic digestion — a reliable and effective process for sludge treatment at small works. *Water Science & Technology*, **22**(3/4), 225–232.

Naegel, L. C. A. (1990) A review of public health problems associated with the integration of animal husbandry and aquaculture, with emphasis on South-East Asia. *Biological Wastes*, **31**(1), 69–83.

Nelson, G. S. (1988) Parasitic zoonoses. *The Biology of Parasitism*, eds Englund, P. T. & Sher, A., New York, Alan R. Liss, Inc., 13–41.

Newman, P. J., Bowden, A. V. & Bruce, A. M. (1989) Production, treatment and handling of sewage sludge. *Sewage Sludge Treatment and Use*, eds Dirkzwager, A. H. & L'Hermite, P., London, Elsevier Applied Science, 11–38.

Newton, W. L., Bennet, H. J. & Foggot, W. E. (1949) Observations on the effect of various sewage treatment processes upon the eggs of *Taenia saginata*. *American J. Hygiene*, **49**, 166–175.

Niederwöhrmeier, B., Böhm, R. & Strauch, D. (1985) Microwave treatment as an alternative pasteurisation process for the disinfection of sewage sludge — experiences with the treatment of liquid manure. In *Inactivation of Microorganisms in Sewage Sludge by Stabilisation Processes*, eds Strauch, D., Havelaar, A. H. & L'Hermite, P., London, Elsevier Applied Science Publishers, 135–147.

Ottolenghi, A. C. & Hamparian, V. V. (1987) Multiyear study of sludge application to farmland: prevalence of bacterial enteric pathogens and antibody status of farm families. *Appl. & Environm. Microbiol.*, **53**(5), 1118–1124.

Owen, J. E. & Stansfield, J. M. (1989) Approaches to safe disposal of farm sludge and slurry. Symposium on *Water Pollution: Microbiology, Chemistry and Treatment*, Birmingham University, 27–28 September.

Owen, R. R. (1983) The effectiveness of chemical disinfection on parasites in sludge. Water Research Centre Conference on *Stabilisation and Disinfection of Sewage Sludge*, Manchester, UMIST, 12–15th April.

Owen, R. R. (1984) The effectiveness of chemical disinfection on parasites in sludge. *Sewage Sludge Stabilization and Disinfection*, ed. Bruce, A. M., Chichester, Ellis Horwood, 426–439.

Owen, R. R. & Crewe, W. (1985) The effect of chemical disinfectants on *Taenia* eggs. In *Inactivation of Microorganisms in Sewage Sludge by Stabilisation Processes*, eds Strauch, D., Havelaar, A. H. & L'Hermite, P., London, Elsevier Applied Science Publishers, 148–157.

Panicker, P. V. R. C. & Krishnamoorthi, K. P. (1981) Parasite egg and cyst reduction in oxidation ditches and aerated lagoons. *J. Water Polln. Control Fed.*, **53**(9), 1413–1419.

Parent, B. C. (1983) Composting technical and economic aspects. *Disinfection of Sewage Sludge: Technical, Economic and Microbiological Aspects*, eds Bruce, A. M., Havelaar, A. H. & L'Hermite, P., Dordrecht & London, Reidel, 139–149.

Paulsrud, B. & Eikum, A. S. (1984) Experiences with lime stabilisation and composting of sewage sludge. Ch. 13 in *Sewage Sludge Stabilisation and Disinfection*, ed. Bruce, A. M., Chichester, Ellis Horwood, 261–277.

Peeters, J. E., Arez Mazás, E., Masschelein, W. J., Villacorta, M. de M. I. & Debacker, E. (1989) Effect of disinfection of drinking water with ozone or chlorine dioxide on survival of *Cryptosporidium parvum* oocysts. *Appl. & Environm. Microbiol.*, **55**(6), 1519–1522.

Pfefferkorn, E. R. (1988) *Toxoplasma gondii* viewed from a virological perspective. *The Biology of Parasitism*, eds Englund, P. T. & Sher, A., New York, Alan R. Liss, Inc., 479–501.

Pfuderer, G. (1985) Influence of lime treatment of raw sludge on the survival of pathogens, on the digestibility of sludge and on the production of methane — hygienic investigations, *Inactivation of Microorganisms in Sewage Sludge by Stabilisation Processes*, eds Strauch, D., Havelaar, A. H. & L'Hermite, P., London, Elsevier Applied Science Publishers, 85–97.

Philipp, W. (1981) Vergleichende hygienische Untersuchungen über die Wirkung der Klärschlammpasteurisierung vor und nach der mesophilen, anaeroben, alkalischen Schlammfaulung. Dissertation Justus Liebig, Universität Giessen.

Philipp, W., Lang, A. & Strauch, D. (1985) Bacteriological and parasitological investigations on the influence of filter beds covered with reed on the survival of *Salmonellas* and *Ascaris* eggs. In *Inactivation of Microorganisms in Sewage Sludge by Stabilisation Processes*, eds Strauch, D., Havelaar, A. H. & L'Hermite, P., London, Elsevier Applied Science Publishers, 168–188.

Pickup, R. W., Saunders, J. R., Morgan, J. A. W., Winstanley, C., Jones, J. G., Carter, J. A., Simon, B. M. & Raitt, F. C. (1989) The fate of genetically engineered microorganisms in freshwater. *WATERSHED '89: The Future for Water Quality in Europe*, eds Wheeler, D., Richardson, M. L. & Bridges, J., IAWPRC/Pergamon, **2**, 375–380.

Pietri, Ch., Hughes, B., Crance, J. M., Puel, D., Cini, C. & Deloince, R. (1988) Hepatitis A virus levels in shellfish exposed in a natural marine environment to the effluent from a treated sewage outfall, *Water Science & Technology*, **20**(11/12), 229–234.

Pike, E. B. (1990) The removal of cryptosporidial oocysts during sewage treatment. In Cryptosporidium *in Water Supplies*, Department of the Environment & Department of Health, London, HMSO, 205–208.

Pike, E. B. & Carrington, E. G. (1986) Inactivation of parasitic ova during disinfection and stabilisation of sludge. *Processing and Use of Organic Sludge and Liquid Agricultural Wastes*, ed. L'Hermite, P., Dordrecht, Reidel, 198–214.

Pike, E. B. & Davis, R. D. (1984) Stabilisation and disinfection — their relevance to agricultural utilisation of sludge. Ch. 3 in *Sewage Sludge Stabilisation and Disinfection*, ed. Bruce, A. M., Chichester, Ellis Horwood, 61–84.

Pike, E. B., Morris, D. L. & Carrington, E. G. (1983) Inactivation of ova of the parasites *Taenia saginata* and *Ascaris suum* during heated anaerobic digestion. *Water Pollution Control*, **82**(4), 501–509.

Pike, E. B., Carrington, E. G. & Harman, S. A. (1988) Destruction of Salmonellas, enteroviruses and ova of parasites in wastewater sludge by pasteurisation and anaerobic digestion. *Water Science & Technology*, **20**(11/12), 337–343.

Poole, J. E. P. & Mills, M. J. (1984) The effect of lime treatment of sewage sludge on the hatching and viability of human beef tapeworn. Ch. 24 in *Sewage Sludge Stabilisation and Disinfection*, ed. Bruce, A. M., Chichester, Ellis Horwood, 440–448.

Quinn, J. J. & Hall, J. E. (1984) Lime stabilisation of undigested sewage sludge. *Sewage Sludge Stabilisation and Disinfection*, ed. Bruce, A. M., Chichester, Ellis Horwood, 540–546.

Rao, V. C. & Melnick, J. L. (1986) *Environmental Virology*, American Society for Microbiology.

Rao, V. C., Metcalf, T. G. & Melnick, J. L. (1987) Removal of indigenous rotaviruses during primary settling and activated-sludge treatment of raw sewage. *Water Research*, **21**(2), 171–177.

Rao, V. C., Metcalf, T. G. & Melnick, J. L. (1988) Recovery of naturally occurring rotaviruses during sewage treatment. *Chemical and Biological Characterisation of Sludges, Sediments, Dredge Spoils and Drilling Muds*, eds Lichtenberg, J. J., Winter, J. A., Weber, C. I. & Fradkin, L., ASTM *STP 976*, 282–287.

Ratcliffe, L. R. (1968) Hatching of *Dicrocoelium lanceolatum* eggs. *Exp. Parasitology*, **23**, 67–68.

Rawcliffe, S. A. & Ollerenshaw, C. B. (1960) Observations on the egg of *Fasciola hepatica*. *Ann. Tropical Med. & Parasitology*, **54**, 172–181.

Reasoner, D. J. (1988) Drinking water microbiology research in the United States: an overview of the past decade. *Water Science & Technology*, **20**(11/12), 101–107.

Reilly, W. J., Collier, P. W. & Forbes, G. I. (1981) *Cysticercus bovis* surveilance — an interim report. *Communicable Disease, Scotland*, 30th May (CDS 81/82), VII–VIII.

Reimers, R. S., Little, M. D., Englander, A. J., Leftwich, D. B., Bowman, D. D. & Wilkinson, R. F. (1982) Parasites in southern sludges and disinfection by standard sludge treatment. *Water Science & Technology*, **14**, 1568.

Reyes, W. L., Krusé, C. W. & Batson, M. St. C. (1963) The effect of aerobic and anaerobic digestion on eggs of *Ascaris lumbricoides* var. *suum* in night soil. *Am. J. Tropical Med.*, **12**, 45–55.

Rickloff, J. R. (1987) An evaluation of the sporicidal activity of ozone. *Applied & Environmental Microbiol.*, **53**(4), 683–686.

Roberts, F. C. (1935) Experiments with sewage farming in south-west United States. *American J. Public Health*, **25**, 122–125.

Rose, J. B., Cifrino, A., Madore, M. S., Gerba, C. P., Sterling, C. R. & Arrowood, M. J. (1986) Detection of *Cryptosporidium* from wastewater and freshwater environments. *Water Science & Technology*, **18**(10), 233–239.

Rose, J. B., Darbin, H. & Gerba, C. P. (1988) Correlations of the protozoa, *Cryptosporidium* and *Giardia*, with water quality variables in a watershed. *Water Science & Technology*, **20**(11/12), 271–276.

Rudd, T. & Hopkinson, L. M. (1989) Comparison of disinfection techniques for sewage and sewage effluents. *J. Instn. Water & Environmental Management*, **3**, 612–617.

Saier, M., Koch, K. & Wekerle, J. (1985) Influence of thermophilic anaerobic digestion (55°C) and subsequent mesophilic digestion of sludge on the survival of viruses without and with pasteurisation of the digested sludge. In *Inactivation of Microorganisms in Sewage Sludge by Stabilisation Processes*, eds Strauch, D., Havelaar, A. H. & L'Hermite, P., London, Elsevier Applied Science Publishers, 28–37.

Scheuerman, P. R., Farrah, S. R. & Bitton, G. (1986) Reduction of microbial indicators and viruses in a cypress strand. In *Health-Related Water Microbiology 1986*, ed. Grabow, W. O. K., *Water Science & Technology*, **18**(10), 1–8.

Schuh, R., Phillipp, W. & Strauch, D. (1985) Influence of sewage sludge with and without lime treatment on the development of *Ascaris suum* eggs. *Inactivation of Microorganisms in Sewage Sludge by Stabilisation Processes*, eds Strauch, D., Havelaar, A. H. & L'Hermite, P., London, Elsevier Applied Science Publishers, 100–110.

Schultz, M. G. (1974) The surveillance of parasitic diseases in the USA. *Amer. J. Tropical Medicine & Hygiene*, **23**, 744–751.

Schwartzbrod, J., Mathieu, C., Thévenot, M. T., Baradel, J. M. & Schwartzbrod, L. (1987) Wastewater sludge: parasitological and virological contamination. In *Use of Soil for Treatment and Final Disposal of Effluents and Sludge*, eds Oliveira, P. R. C. & Almeida, S. A. S., *Water Science & Technology*, **19**(8), 33–40.

Shimohara, E., Sugishima, S. & Kaneko, M. (1985) Virus removal by activated sludge treatment. *Water Science & Technology*, **17**(10), 153–158.

Shuval, H. I., Fattal, B. & Bercovier, H. (1988) Legionnaires diseases and the water environment in Israel. *Water Science & Technology*, **20**(11/12), 33–38.

Silverman, P. H. (1954) Studies on the biology of some tapeworms of the genus *Taenia*. I — factors affecting hatching and activation of their viability. *Ann. Tropical Med. & Parasitology*, **48**, 207–215.

Silverman, P. H. (1955) The survival of the egg of the 'beef tapeworm', *Taenia saginata*. *The Advancement of Science*, **12**(45), 108–111.

Silverman, P. H. & Griffiths, R. B. (1955) A review of methods of sewage disposal in Great Britain with special reference to the epizootiology of *Cysticercus bovis*. *Ann. Tropical Med. & Parasitology*, **49**, 436–450.

Sivinski, H. D. (1975) Treatment of sewage sludge with combinations of heat and ionizing radiation (thermoradiation). *Radiation for a Clean Environment*, I.A.E.A. symposium 17–21 March.

Skirrow, M. B. (1989) *Campylobacter* and the water industry. Symposium on *Water Pollution: Microbiology, Chemistry and Treatment*, Birmingham University, 27–28 September.

Slade, P. J., Fistrovici, E. C. & Collins-Thompson, D. L. (1989) Persistence at source of *Listeria* spp. in raw milk. *Int. J. Food Microbiol.*, **9**, 197–203.

Smith, D. E. (1989) Physico-chemical treatment of wastewater — experience and future development. *WATERSHED '89: The Future for Water Quality in Europe*, eds Wheeler, D., Richardson, M. L. & Bridges, J., IAWPRC/Pergamon, **2**, 307–318.

Smith, H. V. & Rose, J. B. (1990) Waterborne cryptosporidiosis. *Parasitology Today*, **6**(1), reproduced in Cryprosporidium *in Water Supplies*, Department of the Environment and Department of Health, London, HMSO, 146–153.

Smith, M. A., Shah, N. R., Lobel, J. S. & Hamilton, W. (1988) Methaemoglobinaemia and homolytic anaemia associated with *Campylobacter jejuni* enteritis. *Amer. J. Pediatr, Hemat/Oncol.*, **10**, 35–38.

Smyth, J. D. (1976) *Introduction to Animal Parasitology*, 2nd Edition, London, Hodder & Stoughton.

Snow, J. (1855) On the mode of communication of cholera. *Snow on Cholera*, 2nd edition, New York, The Commonwealth Fund.

Snowdon, J. A., Oliver, D. O. & Converse, J. A. (1989a) Land disposal of mixed human and animal wastes: a review. *Waste Management & Research*, **7**, 121–134.

Snowdon, J. A., Oliver, D. O. & Converse, J. A. (1989b) Inactivation of poliovirus 1, as a function of temperature, in mixed human and dairy animal wastes. *Waste Management & Research*, **7**, 135–142.

Sobsey, M. D., Dean, C. H., Knuckles, M. E. & Wagner, R. A. (1980) Intereactions and survival of enteric viruses in soil materials, *Applied & Environmental Microbiology*, **40**, 92–101.

Sobsey, M. D., Shields, P. A., Hauchman, F. H., Hazard, R. L. & Caton, L. W. (1986) Survival and transport of hepatitis A virus in soils, groundwater and wastewater, *Water Science & Technology*, **18**(10), 97–106.

Sobsey, M. D., Fuji, T. & Shields, P. A. (1988) Inactivation of hepatitis A virus and model viruses in water by free chlorine and monochloramine, *Water Science & Technology*, **20**(11/12), 385–391.

Spillmann, S. K., Traub, F., Schwyzer, M. & Wyler, R. (1987) Inactivation of animal viruses during sewage sludge treatment, *Appl. & Environm. Microbiol.*, **53**(9), 2077–2081.

Stetzenbach, L. D., Arrowood, M. J., Marshall, M. M. & Sterling, C. R. (1988) Monoclonal antibody based immunofluorescent assay for *Giardia* and *Cryptosporidium* detection in water samples. *Water Science & Technology*, **20**(11/12), 193–198.

Stien, J. L. & Schwartzbrod, J. (1990) Experimental contamination of vegetables with helminth eggs. *Water Science & Technology*, **22**(9), 51–57.

Stone, A. (1977) Cyst nematodes — most successful parasites. *New Scientist*, 10th November, 355–356.

Storey, G. W. (1987) Survival of tapeworm eggs, free and in proglottids, during simulated sewage treatment processes. *Water Research*, **21**(2), 199–203.

Strauch, D. (1983) Hygienic aspects of the composting process. *Disinfection of Sewage Sludge: Technical, Economic and Microbiological Aspects*, eds Bruce, A. M., Havelaar, A. H., & L'Hermite, P., Dordrecht & London, Reidel, 150–164.

Strauch, D. (1989) Improvement of the quality of sewage sludge. *Sewage Sludge Treatment and Use*, eds Dirkzwager, A. H. & L'Hermite, P., London, Elsevier Applied Science, 160–179.

Strauch, D. & Berg, T. (1980) Mikrobiologische Untersuchungen zur Hygienisierung von Klärschlamm. 2: Versuche an Pasteurisierungssanlagen. *GWF-Wasser/Abwasser*, **121**, 184–187.

Strauch, D., Berg, T. & Fleischle, W. (1980a) Mikrobiologische Untersuchungen zur Hygienisierung von Klärschlamm. 4: Untersuchungen an Bioreaktoren. *GWF-Wasser/Abwasser*, **121**, 331–334.

Strauch, D., Berg, T. & Fleischle, W. (1980b) Mikrobiologische Untersuchungen zur Hygienisierung von Klärschlamm. 5: Untersuchungen an Bio-Zellen-Reaktoren. *GWF-Wasser/Abwasser*, **121**, 392–394.

Strauch, D. & de Bertoldi, M. (1986) Microbiological specifications of disinfected sewage sludge. *Processing and Use of Organic Sludge and Liquid Agricultural Wastes*, ed. L'Hermite, P., Dordrecht, Reidel, 178–197.

Strauch, D., Hammel, H.-E. & Philipp, W. (1985) Investigations on the hygienic effect of single stage and two-stage aerobic-thermophilic stabilisation of liquid raw sludge. *Inactivation of Microorganisms in Sewage Sludge by Stabilisation Processes*, eds Strauch, D., Havelaar, A. H. & L'Hermite, P., London, Elsevier Applied Science Publishers, 48–63.

Strauch, D., Havelaar, A. H. & L'Hermite, P. (eds) (1985) *Inactivation of Microorganisms in Sewage Sludge by Stabilisation Processes*, London, Elsevier Applied Science Publishers.

Telzak, E. E., Bell, E. P., Kautter, D. A., Crowell, L., Budnick, L. D., Morse, D. L. & Schultz, S. (1990) An international outbreak of type E botulism due to uneviscerated fish. *J. Infectious Diseases*, **161**, 340–342.

Theis, J. H., Bolton, V. & Storm, D. R. (1978) Helminth ova in soil and sludge from twelve US urban areas. *J. Water Polln. Control Fed.*, **50**(11), 2485–2493.

Thévenot, M. Y., Larbaigt, G., Collomb, J., Bernard, C. & Schwartzbrod, J. (1985) Recovery of helminth eggs in compost in the course of composting. In *Inactivation of Microorganisms in Sewage Sludge by Stabilisation Processes*, eds Strauch, D., Havelaar, A. H. & L'Hermite, P., London, Elsevier Applied Science Publishers, 158–167.

Vaughn, J. M., Chen, Y. S., Lindburg, K. & Morales, D. (1978) Inactivation of human and simian rotaviruses by ozone. *Appl. & Environm. Microbiol.*, **53**(9), 2218–2221.

Versteegh, J. F. M., Havelaar, A. H., Hoekstra, A. C. & Visser, A. (1989) Complexing of copper in drinking water samples to enhance recovery of *Aeromonas* and other bacteria. *J. Appl. Bact.*, **67**, 561–566.

Voorburg, J. H. & van den Berg, J. J. (1989) Odour problems with sewage sludge. In *Sewage Sludge Treatment and Use*, eds Dirkzwager, A. H. & L'Hermite, P., London, Elsevier Applied Science, 102–115.

Waite, T. D., Kurucz, C. N., Cooper, W. J., Narbaitz, R. & Greenfield, J. (1989) Disinfection of wastewater effluents with electron radiation. In *Environmental Engineering*, Proc. 1989 Specialty Conference, Austin, Texas, ed. Malina, J. F., American Society of Civil Engineers, 619–627.

Warhurst, D. C. (1989) Update on amoebae and *Giardia*. Symposium on *Water Pollution: Microbiology, Chemistry and Treatment*, Birmingham University, 27–28 September.

Warrell, D. A. (1990) Clinical features of cryptosporidiosis. Cryptosporidium *in Water Supplies*, Department of the Environment & Department of Health, London, HMSO, 141–153.

Warren, K. S. (1988) The global impact of parasitic diseases. *The Biology of Parasitism*, eds Englund, P. T. & Sher, A., New York, Alan R. Liss, Inc., 3–12.

Watson, D. C., Satchwell, M. & Jones, C. E. (1983) A study of the prevalence of parasitic helminth eggs and cysts in sewage sludges disposed to agricultural land. *Water Polln. Control*, **82**(3), 285–289.

Webbe, G. (1967) The hatching and activation of Taeniid ova in relation to the development of cysticercosis in man. *Zeitschrift für Tropenmedezin und Parasitologie*, **19**, 354–369.

Wekerle, J., Leuze, M. & Koch, K. (1985) Virus inactivating effect of anaerobic mesophilic digestion of municipal sludge with or without different preceding types of treatment. *Water Science & Technology*, **17**(10), 175–184.

Wharton, D. A. (1979) The structure and formation of the egg-shell of *Syphacia obvelata* Rudolphi (Nematoda: Oxyuroidea). *Parasitology*, **79**, 13–28.

Wheeler, D., Richardson, M. L. & Bridges, J. (1989) *WATERSHED '89: The Future for Water Quality in Europe*, **2**, IAWPRC/Pergamon.

White, K. E. (1984) Ionising radiation to treat sludge: energy sources and future prospects. Ch. 26 in *Sewage Sludge Stabilisation and Disinfection*, ed. Bruce, A. M., Chichester, Ellis Horwood, 462–486.

Wiley, B. B. & Westerberg, S. C. (1969) Survival of human pathogens in composted sewage. *Appl. Microbiology*, **18**(6), 994–1001.

Wilkens, S. (1982) Risk of infection of cattle by *Taenia saginata* and *Sarcocystis* spp. from irrigated sewage plant effluents and *in vitro* trials on the rate of sedimentation of helminth eggs. *Veterinary Bulletin*, **52**, 2081 (cited by Snowdon *et al.*, 1989a).

Williams, D. E., Elder, D. & Worley, S. D. (1988) Is free halogen necessary for disinfection? *Applied & Environmental Microbiology*, **54**(10), 2583–2585.

Williams, P. T. (1990) A review of pollution from waste incineration. *J. Inst. Water & Environmental Management*, **4**, 26–34.

Wilson, R. A. (1967) The structure and permeability of the shell and vitelline membrane of the egg of *Fasciola hepatica*. *Parasitology*, **57**, 47–58.

Winkler, M. A. (1981) *Biological Treatment of Waste-Water*, Chichester, Ellis Horwood.

Winkler, M. A. (1984) Biological control of nitrogenous pollution in wastewater. Ch. 3 in *Topics in Enzyme and Fermentation Biotechnology*, ed. Wiseman, A., Chichester, Ellis Horwood, **8**, 31–124.

Winkler, M. A. & Manoranjan, V. S. (1989) Control of nitrogenous pollution. In *WATERSHED '89: The Future for Water Quality in Europe*, eds Wheeler, D., Richardson, M. L. & Bridges, J., IAWPRC/Pergamon, **2**, 463–473.

Wizigmann, I. & Würschung, F. (1974) Experience with a pilot plant for irradiation of sewage sludge. Proc. I.A.E.A. summer meeting.

Wright, A. E. (1990) *Cryptosporidium* — the organism. In Cryptosporidium *in Water Supplies*, Department of the Environment & Department of Health, London, HMSO, 77–81.

Wright, W. H., Cram, E. B. & Nolan, M. O. (1942) Preliminary observations on the effect of sewage treatment processes on ova and cysts of intestinal parasites. *Sewage Works J.*, **14**, 1274–1280.

Yates, M. V. & Yates, S. R. (1988) Virus survival and transport in ground water. *Water Science & Technology*, **20**(11/12), 301–307.

Young, C. E. & Carlson, G. A. (1975) Land treatment *versus* conventional advanced treatment of municipal wastewater, *Water Pollution Control*, **47**(11), 2565–2573.

Index